About Island Press

Island Press is the only nonprofit organization in the United States whose principal purpose is the publication of books on environmental issues and natural resource management. We provide solutions-oriented information to professionals, public officials, business and community leaders, and concerned citizens who are shaping responses to environmental problems.

In 2006, Island Press celebrates its twenty-second anniversary as the leading provider of timely and practical books that take a multidisciplinary approach to critical environmental concerns. Our growing list of titles reflects our commitment to bringing the best of an expanding body of literature to the environmental community throughout North America and the world.

Support for Island Press is provided by the Agua Fund, The Geraldine R. Dodge Foundation, Doris Duke Charitable Foundation, The William and Flora Hewlett Foundation, Kendeda Sustainability Fund of the Tides Foundation, Forrest C. Lattner Foundation, The Henry Luce Foundation, The John D. and Catherine T. MacArthur Foundation, The Marisla Foundation, The Andrew W. Mellon Foundation, Gordon and Betty Moore Foundation, The Curtis and Edith Munson Foundation, Oak Foundation, The Overbrook Foundation, The David and Lucile Packard Foundation, The Winslow Foundation, and other generous donors.

The opinions expressed in this book are those of the author(s) and do not necessarily reflect the views of these foundations.

Conservation Across Borders

CONSERVATION ACROSS BORDERS

Biodiversity in an Interdependent World

Charles C. Chester

WASHINGTON • COVELO • LONDON

Copyright © 2006 Charles C. Chester

All rights reserved under International and Pan-American Copyright Conventions. No part of this book may be reproduced in any form or by any means without permission in writing from the publisher: Island Press, 1718 Connecticut Ave., NW, Suite 300, Washington, DC 20009.

ISLAND PRESS is a trademark of The Center for Resource Economics.

The frontispiece quotations are from Kurkpatrick Dorsey, *The Dawn of Conservation Diplomacy* (1998, Seattle: University of Washington Press, page 4) and Peter Goodwin, "Without Borders" (*National Geographic*, September 2001).

Figure 2.2 is used with permission from *Endangered Species UPDATE*; Figure 3.1 is used with permission from the International Sonoran Desert Alliance; and Figure 4.1 is used with permission from the Yellowstone to Yukon Conservation Initiative.

Library of Congress Cataloging-in-Publication Data
Chester, Charles C.
  Conservation across borders: biodiversity in an interdependent world / Charles C. Chester
     p.cm.
  Includes bibliographical references (p.   ).
  ISBN 1-55963-610-6 (cloth : alk. paper) — ISBN 1-55963-611-4 (pbk. : alk. paper)
  1. Biological diversity conservation—North America—International cooperation.   I. Title.
QH77.N56C48 2006
333.72—dc22   2005031591

Printed on recycled, acid-free paper ♻

Design by Joan Wolbier

Manufactured in the United States of America
10 9 8 7 6 5 4 3 2 1

For Lael, Sam, and Caleb

*It is evident that natural resources
are not limited by the boundary lines
which separate nations, and that the need
for conserving them upon this continent is
as wide as the area upon which they exist.*
—Theodore Roosevelt

Political boundaries are the scars of history.
—Willem van Riet, Peace Parks Foundation

CONTENTS

Preface XIII

1 Rising Above the Territories of Chance 1

2 The Multiple Personalities of Transborder Conservation 14

3 Border Biosphere Reserves and the International Sonoran Desert Alliance 53

4 Landscape Vision and the Yellowstone to Yukon Conservation Initiative 134

5 Conservation Effectiveness in the Territories of Chance 217

Acknowledgments 241

List of Interviewees 245

List of Abbreviations 247

A Note on Web Notes 249

Index 251

PREFACE

In the mid-1990s, I was a Volunteer in the Park—a VIP, as they generously called us—in Yellowstone National Park. My assignment concerned management of the highly diverse extremophilic microorganisms that live in the 10,000 or so thermal features scattered throughout the park. At the time, I recognized the irony of working on microscopic critters in the largest national park of the lower forty-eight (I hadn't yet heard of Death Valley's 1994 upgrade from a monument to a park, superceding Yellowstone's geographic preeminence). Little did I know that I would spend the subsequent decade studying landscapes that would dwarf Yellowstone National Park.

Also during that period, I had enrolled in the Ph.D. program at The Fletcher School at Tufts University. At a 1996 meeting of the Greater Yellowstone Coalition, Professor Tim Clark of Yale University suggested that I use my dissertation as an opportunity to look at the inchoate Yellowstone to Yukon Biodiversity Strategy (Y2Y). I immediately grabbed on to the idea. Over the following year, with the goal of conducting a large-scale comparative analysis, I spent a good deal of time drawing up a long list of other transborder conservation initiatives on the North American continent. Once I was disabused of the ludicrous notion of analyzing them all, I dramatically narrowed my scope to compare Y2Y with the International Sonoran Desert Alliance (ISDA).

Despite significant differences, both ISDA and Y2Y emanated from civil society. My essential question for both was this: How might the respective monikers of ISDA and Y2Y enhance the ability of their participants to "do" conservation in the real world? In the parlance of contemporary conservation circles, I was interested in their *effectiveness* in promoting on-the-ground conservation. I dissected that broad issue into a long list of specific questions, resulting in a standardized interview that I would eventually use with sixty participants in ISDA and Y2Y. (See "List of Interviewees.") Wherever I have quoted an individual without providing a parenthetical citation in this book, the quote comes from one of these interviews.

Many wise people warned me to keep my interview short. Ignoring all of them, I ultimately paid the price by spending several months coding the resultant transcripts with the qualitative analysis software package NUD*IST—one of my favorite acronyms, which stands for Non-numerical, Unstructured Data: Indexing, Searching, and Theorizing. (Upgrades to this software package have unfortunately been renamed, and there are now several competitive packages on the market.) I strongly recommend the use of such an aid—without NUD*IST, I'd probably still be sequestered in some library carrel, buried in index cards.

From the perspective of a qualitative researcher, what surprised me the most was how differently the Y2Y participants and the ISDA participants interpreted the same questions. As will become apparent in reading the two case study chapters, ISDA participants responded to my questions in terms of a particular *history*, whereas Y2Y participants responded by describing a particular *idea*. Were I to start over on this project, I might tighten my questions to minimize room for such interpretation. But maybe not; as I describe in Chapter 5, the differing interpretations may offer the most useful lesson for conservationists in other transborder landscapes.

Ultimately, this book constitutes a condensation and update of my dissertation, completed in 2003. In addition to proving itself an effective doorstop, the dissertation contains a good deal more information on ISDA and Y2Y—as well as an analysis of how they fit into various theoretical frameworks expounded in the field of international relations. Most readers, I expect, will be happy to know that the editors at Island Press reined in my pack-rat mentality, thereby ensuring that the book focuses on key issues related to transborder conservation. However, for interested readers, some of that information is available in the form of an extensive set of "Web notes" to the book at www.islandpress.org/conservation_across_borders/webnotes.pdf.

My greatest hope and fear for this book is identical: that it may be seen as telling *the* story of ISDA or *the* story of Y2Y. But whereas it is my hope for cause of ego, it is my fear for cause of omission. Through my research over the past years, ISDA and Y2Y have been my constant companions (and just like human companions, they have been a source of enduring happiness and intermittent annoyance). But a mere glance through my files—or procrastination at my bookshelves—never fails to reveal another character, another event, or another theme that I've missed in my narratives. So I have come to know ISDA and Y2Y well enough to understand that this book only scratches the surface of the stories they have to tell. Ultimately, *the* stories of ISDA and Y2Y cannot now be told; both are, praise be, living institutions whose stories have hopefully just begun.

Charles C. Chester
November 2005

1

Rising Above the Territories of Chance

Over three hundred years before North America's borders took their modern shape, a Spanish explorer named Álvar Núñez Cabeza de Vaca became stranded on the western shores of *la Florida*. For nearly a decade, he and his three companions (one a black slave) did not so much travel as *live* across the present-day states of Texas, Chihuahua, Sonora, and Sinaloa—finally encountering "Christians" near the Pacific coast in 1536. Retold in countless narratives about Spanish colonization of the American Southwest, their story resonates because they were the first Europeans—and one African—to cross the continent starting from a point within the present-day United States. But more strikingly, it resonates because of their resilience across a vast landscape of which they were utterly ignorant (Cabeza de Vaca 1555).

Cabeza de Vaca's terse account of his odyssey reveals a man who respected the indigenous cultures in which he immersed himself, and who recognized the land he crossed as a complex mosaic of biological and cultural patterns. The idea of a borderline cutting across that landscape, a line demarcating vastly different land-use practices and cultures to either side, would likely have been inconceivable to him. So although we do not know exactly when or where he and his companions stepped over the nonexistent international border, we can safely assume that the event was marked by all the pomp and circumstance of just another footstep.

Although borders often conform in some humanly logical way to the lay of the land (think of large rivers and mountain ranges), in the western portion of North America we have strewn straight-line borders across the landscape on the basis of diplomatic (and not so diplomatic) decisions born of geographical ignorance (see Carroll 2001; Rebert 2001). In so doing, we have unwittingly granted borders the power to carve out the fortified nation-states—or what I describe as the *territories of chance*—that so strongly define who we are. "I am a Canadian," one says if born and bred in Vancouver, and similar phrases can be heard in San Diego and Tijuana. Yet in each case, it is possible to imagine the historical dice rolling a different way.

Unfolding in these territories of chance are the two principal case studies of this book, the International Sonoran Desert Alliance (ISDA) and the Yellowstone to Yukon Conservation Initiative (Y2Y). In the case of ISDA, if the United States had not acquired land from Mexico under the Gadsden Purchase of 1853–54, the area of concern would have been nearly entirely within Mexico—with the somewhat ironic result that the border would have more closely demarcated ecosystem "boundaries." As to Y2Y, the 49th parallel divided the region due to little more than the compromises of geographic ignorance under the Treaties of 1818 and 1846 between the United States and Great Britain (for Canada).

So in one sense, these international borders are little more than unerringly straight one-dimensional lines that define two-dimensional areas on a map. Yet on the ground, borders often create severe cultural, political, and biological effects—even where a border reflects something of the land's true character. To get a sense of a border's physical impact, one need only stand on the Bridge of the Americas between El Paso, Texas, and Ciudad Juarez, Chihuahua. That vantage point affords the dismal sight of an inanimate, canalized river "entombed" in concrete, and one can hardly imagine the former Río Bravo—Rio Grande for gringos—that once teemed with life (Weisman and Dusard 1986, 100). Or consider the largely regrown forests stretching across the northernmost parts of New York and New England, which in the United States have collectively come to be called "the Northern Forest." That name makes little sense to the average Quebecer whose ancestors largely deforested the landscape immediately north of the international border. Their true "northern forest" is the boreal (the tendrils of which reach to within a two-hour drive of Ottawa), and not the tracts of Adirondack or New England forest that they drive through on southward treks to New York or Boston (Gawthrop 1999). Overall, the fact that immaterial one-dimensional lines can have such tangible multidimensional effects is nothing less than extraordinary. But we are so used to borders that it is difficult to see them—or their territories of chance—as anything but normal.

As their individual stories will show, ISDA and Y2Y are attempting to redefine the territories of chance by replacing them with a "geography of hope," as Wallace Stegner (1960) so quotably put it. If nothing else, they are trying to rise above their territories of chance—to overcome the unnatural divisions in their regions by making some kind of sense out of the living landscapes they inhabit.

International Conservation: Complicating Matters to Solve a Problem

Against the trends of human population growth, increasing resource consumption, and the rapid pace of technological change, governments and conservation organizations have protected biodiversity by implementing various types of domestic conservation programs. Yet such initiatives in the domestic realm have repeatedly proven to become contentious and strife-ridden affairs that in countless cases seem to be unresolvable. Given the obstacles surrounding biodiversity protection on the domestic front, it is reasonable to wonder why one would choose to focus on the more difficult problem of international conservation. As Westing put it (1998, 91), working for conservation "with two (or occasionally even three) sovereign states involved would seem to add a gratuitous layer of complexity that spells almost certain failure. So why try?"

The principal response is obvious: Biodiversity knows no political boundaries. Transborder conservation cannot await "problem resolution" at the domestic level for the simple reason that biodiversity has evolved not in conformance to the dictates of political geography, but rather in accordance with the biogeographic forces of natural selection. If it is true that "approximately one-third of all terrestrial high-biodiversity sites straddle national borders" (Westing 1993, 5; 1998, 91), then effective conservation must take into account this inherent transpolitical nature of biodiversity. In short, although purely domestic approaches to biodiversity conservation have been and will be critical, protecting life on Earth will ultimately require an international approach.

All of this is old hat to practicing conservationists, most of whom know full well that "saving the world" means working across international borders. Indeed, the multiple environmental impacts of borders have elicited a broad range of responses. A 1996 survey of the U.S.-Mexico border environmental activities, for instance, listed 485 distinct transborder initiatives and projects (most of them related to environmental education and pollution control; see U.S.-Mexico Border XXI Program 1996).

Along with the sheer number of initiatives that survey identified, it also revealed a significant development in how transborder conservation takes place. Since the "dawn of conservation diplomacy" at the end of the nineteenth century, individual citizens and nongovernmental organizations (NGOs) have played an integral role in international environmental affairs (Dorsey 1998). Collectively described as "civil society," these nongovernmental, nonmarket actors have traditionally focused on influencing (or reacting to) governmental actions—governments being the traditional arbiter of cross-border cooperation (see DeSombre 2000). More recently, however, civil society actors have taken on the task of reaching across borders to work together directly *without* input from or influence on governmental entities. ISDA and Y2Y are two noteworthy examples of this trend. Under both, individual citizens and NGOs have banded together with the first-order priority of coming to know and understand each other. Through such networking, participants in ISDA and Y2Y have been empowered to focus on their ultimate mission: to protect each region's ecological and cultural cohesiveness from the threat of political boundaries.

It is important to note that although ISDA and Y2Y share this broad mission, the two initiatives commenced in the early 1990s with substantially differing conceptual foundations. ISDA participants concentrated on building an environmentally sound regional economy, with biodiversity conservation as one component. Y2Y's first and foremost issue was biodiversity conservation, although most participants soon came to recognize the need to focus on the dramatic economic changes facing local communities. In essence, participants in both initiatives struggled over what constitutes "sustainable development" and how that loose concept relates to biodiversity conservation. And although the word *biodiversity* was heard far more often in the halls of Y2Y than in those of ISDA, the concept of biodiversity was central to both. It is thus necessary to take a look at the history of what biodiversity has come to mean.

Save the Whales (and Fifteen Thousand Other Species)

Over millennia, evolution and migration have endowed the North American continent with an immense variety of ecosystems, species, and individual populations. This diversity has never been static; since North

America separated from Asia and Europe about 120 million years ago, ecosystems have shifted and transformed, species have come and gone, and individual populations have appeared and disappeared. In the continent now host to the artifact boundaries between Canada, the United States, and Mexico, the only biological constant has been change (Flannery 2001).

Yet even as biologists and ecologists have moved past romantic notions of the "balance of nature," they have discovered that rates of change in biological and ecological processes have accelerated due to the apparently ineluctable absorption of the earth's biological productivity for the use of *Homo sapiens* (Haberl et al. 2004). Within all biomes of the continent, from grasslands and deserts to montane forests and arctic tundra, the biological landscape is experiencing an unprecedented degree of alteration that has led to catastrophic results for many of the continent's species. On a global scale, scientists have focused on the loss of species, pointing out that human activities have increased rates of species extinction by anywhere between 100 to 1,000 percent compared to normal "background rates" (i.e., the rate of extinctions on a geologic timescale not including the five major "mass extinctions" that paleontologists have detected in the fossil record) (Levin and Levin 2002; Regan et al. 2001).

Up until less than two decades ago, scientists and conservationists referred to this general problem with myriad terms and slogans that ranged from "endangered species" and "silent spring" to "Bring back the wolf" and "Save the whales." The latter, perhaps constituting the planet's most widely recognized call to conservation, seems to offer a straightforward message. But it is worth pondering what is actually entailed in that simple catchphrase. For example, there are approximately 78 species of whales, each with its own conservation requirements—some needing a lot of help, some arguably little (Waller 1996, 397). Saving 78 species sounds relatively doable, and, happily, the prospects for at least a few whale species have improved over the past two decades due to international cooperation (Corkeron 2004; Payne 1995, 296). But what about 15,503 species? That is the total number of threatened species listed by the World Conservation Union (IUCN), many of which will also require individualized prescriptions (or proscriptions) in order to be "saved" (World Conservation Union 2004). Even worse, 15,503 is widely recognized as a severe underestimate.

With so many documented and undocumented threatened species, the practical implications of fifteen thousand iterations of "Save the . . ." is overwhelming. So how do we think about this problem comprehensively? One answer came through the widespread adoption of the catchall concept of biological diversity. Although the phrase dates at least to the 1960s (see, for example, Krutilla 1967, 786), its contemporary use is dated to the 1980 publication of two U.S. government reports (Lovejoy 1980; Norse and McManus 1980). Its shorthand version, *biodiversity*, did not become adopted within the conservation community until after the 1986 National Forum on BioDiversity and the publication of its proceedings as the book *Biodiversity*, edited by the well-known Harvard biologist Edward O. Wilson (1988). Use of the term has since dramatically increased; it is now widely employed by physical scientists, conservationists, and social scientists and has been adopted into the common vocabulary of the media (Jeffries 1997, Chapter 1). And regardless of whether the "reading public" has itself become an endangered species, it is worth noting that approximately four hundred commercially available books have the word *biodiversity* in their titles—and that at least at one time, the word generated more finds on Google than did "climate change," "Beatles," "George W. Bush," or "Tiger Woods" (Norse and Carlton 2003).

So what exactly does *biodiversity* mean? A generic and ubiquitous definition is "the totality and variety of life on Earth" (see, for example, UNDP n.d.), but it has been pointed out that that definition is nearly equivalent to the long-extant term *biota* (Vogel 1994, 18). By 1986, however, biologists and ecologists had worked out a more rigorous definition that proposed a three-tiered hierarchy of diversity: the diversity of ecosystems, the diversity of species, and the diversity of individuals and populations within a single species (commonly called "genetic diversity" but more precisely referred to as "allelic diversity") (Norse et al. 1986). Whether writing journal articles or grant applications, scientists today regularly rely on the term *biodiversity* as shorthand for a distinction between these three levels. And when those in the life sciences talk or write about having to save "biodiversity" (which some of them do quite often, others not nearly enough), they typically take care to distinguish between the natural phenomenon of "biodiversity" on the one hand, and the advocacy position of "biodiversity conservation" (or some such pairing) on the other. The shorthand language of "biodiversity" thus makes it easier for scientists to draw attention to the broad anthropogenic forces that threaten the planet's ecosystems, species, and genetic resources.

Yet regardless of its scientific definition, biodiversity is often simply equated with conservation and has been described as the "rallying cry" of the global extinction crisis (Takacs 1996). As is plainly evident from the sixty-two individual contributions to Wilson's pathbreaking book *Biodiversity*—all of which concern the loss of biodiversity—the word has been intimately tied to its imperiled status from the start. Furthermore, it is now widely understood in terms of a social movement—a result due in no small part to political contention over the 1992 Convention on Biological Diversity (CBD), more commonly known as the Biodiversity Convention. Concerned over intellectual property rights for biotechnology, the first Bush administration declined to sign the convention at the 1992 Earth Summit in Rio de Janeiro (President Clinton signed the CBD in 1993, but the U.S. Senate has not ratified it). The refusal to sign provoked clamorous opposition from the conservation community, strong dissent from numerous developed and developing countries, and a good deal of attention from the media, thus giving the word *biodiversity* a cachet it would likely have taken years to garner otherwise.

Ultimately, there is good cause to associate the word with conservation. Scientists and conservationists have grouped the human-caused threats to biodiversity into six general categories: overexploitation, loss of habitat, degradation of habitat, invasive species, climate change, and secondary effects (and multiple permutations thereof—this classification system will be revisited in Chapter 5). These six categories are often described as the direct or "proximate" causes of biodiversity loss, as opposed to the underlying or "root" causes such as human population growth, overconsumption, and technological change. And it is important to keep in mind that these multifarious threats constitute tremendous challenges not only for individual species, but for ecosystems and genetic diversity as well. In the United States, for example, scientists have identified eighty-two ecosystem types that are threatened, endangered, or critically endangered (Noss et al. 1995). At the genetic level, scientists have estimated that individual populations of species are being extirpated at a rate of 16 million per year out of an estimated total of 1.1 to 6.6 billion populations worldwide (see Ehrlich and Daily 1997–98; Hughes et al. 1997). Overall, Wilson has argued that biodiversity loss is the most irreparable environmental challenge facing humanity, with consequences more lasting than ozone depletion, toxic releases, transborder air pollution, and even nuclear war (Christen 2000; Myers 1993, 159). In contrast to those, as a popular saying goes, "Extinction is forever."

The problem of biodiversity loss has been addressed by investigators in a wide range of academic disciplines, including philosophy (Rolston 1994), economics (McNeely 1988), political science (Tobin 1990), international law (Heijnsbergen 1997), history (Reiger 2001), geography (Platt 1996), linguistics (Magurran 1988), policy analysis (Takacs 1996), education (Jacobson 1995), anthropology (Davis 1998), chemistry (Eisner 1994), health (Grifo and Rosenthal 1997), medicine (Aguirre et al. 2002), physics (Gell-Mann 1994), geology (Ward 1994), and, of course, a myriad of subdisciplines in biology (e.g., forestry, ethnobotany, limnology, and genetics). This book looks to many of those disciplinary perspectives to address transborder conservation of biodiversity.

Overview of the Book

Chapter 2 provides a broad overview of the global history and current status of transborder conservation. From the establishment of early "peace parks" to the designation of continental migratory pathways, the chapter sets out the full panoply of transborder mechanisms to protect biodiversity. It then examines the rise of "bioregionalism" and the history of international cooperation under the aegis of UNESCO's Man and the Biosphere (MAB) Program. The chapter ends by arguing that even while many of the conceptual roots of ISDA and Y2Y can be traced to MAB and its global network of "biosphere reserves," both initiatives are significant departures from earlier transborder work.

Chapters 3 and 4 address each of the two case studies respectively with a fundamental question in mind. In a world of endangered species, degraded ecosystems, and burgeoning human populations, have ISDA and Y2Y tangibly enhanced biodiversity conservation and the possibility of sustainable economies? Or to rephrase the question in terms of international cooperation: Has working together across international borders tangibly enhanced the effectiveness of nongovernmental actors in their pursuit of biodiversity conservation and sustainable development? To answer that question, both chapters focus on how the respective initiatives were established, what their overall intentions and functions were, and how they have or have not translated intentions into action.

Chapter 3 takes place within the Sonoran Desert, one of the world's most biologically diverse deserts and considered to be one of North America's least anthropogenically modified ecosystems. Whereas the U.S.

government began protecting large portions of the region in the 1930s, official Mexican efforts to protect the region—particularly the extremely arid volcanic region known as the Pinacate—began only in the late 1970s. During the 1980s, many conservationists and scientists became more vocal in expressing their concern that these separate national efforts inadequately addressed both the human and conservation needs of the region as a whole. They believed that it would be more effective to designate a vast portion of the region as an international biosphere reserve. Although one particular protected area on the U.S. side of the border had been designated a biosphere reserve in 1976, the designation was little more than a label since there had been no implementation of the fundamental biosphere reserve concepts of "core areas" and "buffer zones."

The International Sonoran Desert Alliance was born within this intellectual milieu. This is not to say, however, that ISDA was solely the child of "biosphere reserve thinking"; the numerous individuals who participated in ISDA did so for a variety of reasons, one of the foremost being the lack of indigenous control over traditional lands. Chapter 3 describes both ISDA's complex formation and the various institutional personalities that it took on in subsequent years. One of ISDA's more prominent roles, in addition to advocating for the official designation of an international biosphere reserve, was as a forum for local participation in planning and management of the extant biosphere reserves—the importance of which the international MAB Program had repeatedly emphasized. ISDA's role provides insight into just how complex "local participation" can be, for over the course of the 1990s ISDA faced tremendous obstacles that include internal discord between staff and board members as well as a reactionary political movement aimed at disassociating the United States from international conservation programs. Yet ISDA—as well as the very process of putting ISDA together—made several significant contributions to conservation of the region, including the establishment of a renowned biosphere reserve within Mexico and cooperative agreements both between the states of Arizona and Sonora and between the United States and Mexico.

Chapter 4 examines the Yellowstone to Yukon Conservation Initiative (Y2Y). Several factors led to the synthesis of the essential Y2Y concept of wildlife connectivity throughout the Rocky Mountains that stretch between the United States and Canada—mountains that once blocked development but now beckon a human presence that all too often places insubstantial value on wildlife habitat. While one of these factors was the relatively uniform ecological and biological conditions of the region, equally if not more important factors included general advances in the science of conservation biology, specific findings of conservation biologists regarding the needs of large carnivores and wide-ranging species, and other lessons gleaned from large landscape and transboundary initiatives both within and outside of the Y2Y region.

Initially established as a loose network of conservationists and scientists throughout the region, Y2Y has since become a formal organization. Participants attest that Y2Y has infused the region's conservation community with cooperation, camaraderie, mutual understanding, and inspiration. Once isolated from each other by long distances and narrowly focused agendas, conservationists within the Y2Y region now find themselves working in conjunction with a geographically extensive array of individuals and organizations that have also tied their efforts to the "Y2Y vision." Y2Y's success has largely been attributed to its engaging landscape vision, a vision that has incorporated both missionary and scientific elements within its iconic composition of Yellowstone (the first national park in the world) and the Yukon (the proving ground of Canadian grit, poetry, and national identity).

Y2Y, say its proponents, has the power to engage significant segments of society and to change the way people generally comprehend and care about such an immense region. The conservation community and much of the popular media have portrayed Y2Y as one of the more innovative and promising approaches to conservation on the continent—if not in the world. But all reaction to Y2Y has hardly been favorable. Opponents have portrayed it as an attempt to lock up public lands (or "crown lands," as they have traditionally been called in Canada) and to further regulate private property. Both of those reactions have given the initiative widespread exposure, making Y2Y a relatively well-recognized term in several segments of society. But even if such conceptual shifts toward large-landscape thinking remain limited to those within the conservation community, most participants argue that Y2Y has tangibly enhanced the way conservation is carried out in the region by increasing networking, enhancing funding, and generating more extensive scientific research.

Chapter 5 begins by looking at the various ways conservationists think about and measure the effectiveness of what they do, offering a broad framework of this burgeoning field of "conservation effectiveness." In deference to a widely accepted methodological preference within the social sciences for *comparative* analyses, the chapter then addresses the question of which of the two regional conservation efforts chronicled in this book has been more effective. More specifically, I ask what lessons can be learned about success and failure from a comparative analysis of ISDA and Y2Y. On the face of it, the comparison may seem unfair: ISDA and Y2Y faced dramatically different circumstances in their origins and operations—substantive differences too easily ignored in a simplistic side-by-side comparison. But a side-by-side comparison is not the goal of Chapter 5; rather, it proposes a set of seven specific "independent variables" that relate to conservation effectiveness: mission breadth, constituency inclusion, communication systems, scientific support and participation, leadership dynamics, political backlash, and landscape vision. These independent variables allow us to garner the insights of a comparative analysis without losing sight of the distinct realities facing the two initiatives. The chapter concludes the book with a discussion of the complex relationships among civil society, governments, and international borders.

References

Aguirre, A. Alonso, Richard S. Ostfeld, Gary M. Tabor, Carol House, and Mary C. Pear, eds. 2002. *Conservation medicine: Ecological health in practice.* New York: Oxford University Press.

Cabeza de Vaca, Álvar Núñez. 1555 [1993]. *The account: Álvar Núñez Cabeza de Vaca's Relación.* Houston, TX: Arte Público Press.

Carroll, Francis M. 2001. *A good and wise measure: The search for the Canadian-American boundary, 1783–1842.* Toronto, Canada: University of Toronto Press.

Christen, Kris. 2000. Biodiversity at the crossroads. *Environmental Science & Technology* 34, no. 5: 123A–128A.

Corkeron, Peter J. 2004. Whale watching, iconography, and marine conservation. *Conservation Biology* 18, no. 3: 847–849.

Davis, Wade. 1998. *The clouded leopard: Travels to landscapes of spirit and desire.* Vancouver, Canada: Douglas & McIntyre.

DeSombre, Elizabeth R. 2000. *Domestic sources of international environmental policy: Industry, environmentalists, and U.S. power.* Cambridge, MA: MIT Press.

Dorsey, Kurkpatrick. 1998. *The dawn of conservation diplomacy: U.S.-Canadian wildlife protection treaties in the progressive era.* Seattle: University of Washington Press.

Ehrlich, P. R., and C. G. Daily. 1997–98. Population extinction and the biodiversity crisis. *Wild Earth* 7, no. 4: 35–45.

Eisner, Thomas. 1994. Chemical prospecting: A global imperative. *Proceedings of the American Philosophical Society* 138, no. 3: 385–392.

Flannery, Tim F. 2001. *The eternal frontier: An ecological history of North America and its peoples.* New York: Atlantic Monthly Press.

Gawthrop, Daniel. 1999. *Vanishing halo: Saving the boreal forest.* Vancouver, Canada: Greystone Books.

Gell-Mann, Murray. 1994. *The quark and the jaguar: Adventures in the simple and the complex.* New York: W.H. Freeman.

Grifo, Francesca, and Joshua Rosenthal, eds. 1997. *Biodiversity and human health.* Washington, DC: Island Press.

Haberl, Helmut, Mathis Wackernagel, Fridolin Krausmann, Karl-Heinz Erb, and Chad Monfreda. 2004. Ecological footprints and human appropriation of net primary production: A comparison. *Land Use Policy* 21, no. 3: 279–288.

Heijnsbergen, P. van. 1997. *International legal protection of wild fauna and flora.* Washington, DC: IOS Press.

Hughes, Jennifer B., Gretchen C. Daily, and Paul R. Ehrlich. 1997. Population diversity: Its extent and extinction. *Science* 278, no. 5338: 689–692.

Jacobson, Susan Kay, ed. 1995. *Conserving wildlife: International education and communication approaches.* New York: Columbia University Press.

Jeffries, Mike J. 1997. *Biodiversity and conservation.* New York: Routledge.

Krutilla, John V. 1967. Conservation reconsidered. *American Economic Review* 57, no. 4: 777–786.

Levin, Phillip S., and Donald A. Levin. 2002. The real biodiversity crisis. *American Scientist* 90: 6–8.

Lovejoy, T. E. 1980. Changes in biological diversity. In *The Global 2000 report to the president. Vol. 2: The technical report,* ed. G. O. Barnay, 327–332. Harmondsworth, England: Penguin Books.

Magurran, Anne E. 1988. *Ecological diversity and its measurement.* Princeton, NJ: Princeton University Press.

McNeely, Jeffrey A. 1988. *Economics and biological diversity.* Gland, Switzerland: IUCN.

Myers, Norman. 1993. *Gaia: An atlas of planet management.* London: Gaia Books.

Norse, Elliot A., and Roger E. McManus. 1980. Ecology and living resources: Biological diversity. In *Environmental Quality 1980: The eleventh annual report of the Council on Environmental Quality,* ed. Council on Environmental Quality: 31–80. Washington, DC: U.S. Government Printing Office.

Norse, Elliot A., and James T. Carlton. 2003. World Wide Web buzz about biodiversity. *Conservation Biology* 17, no. 6: 1475–1476.

Norse, Elliot A., Kenneth L. Rosenbaum, David S. Wilcove, Bruce A. Wilcox, William H. Romme, David W. Johnston, and Martha L. Stout. 1986. *Conserving biological diversity in our national forests.* Washington, DC: Wilderness Society.

Noss, R. F., E. T. LaRoe, and J. M. Scott. 1995. *Endangered ecosystems of the United States: A preliminary assessment of loss and degradation.* Biological Report 28. Washington, DC: National Biological Service.

Payne, Roger. 1995. *Among whales.* New York: Scribner.

Platt, Rutherford H. 1996. *Land use and society: Geography, law, and public policy.* Washington, DC: Island Press.

Rebert, Paula. 2001. *La gran línea: Mapping the United States–Mexico boundary, 1849–1857.* Austin: University of Texas Press.

Regan, Helen M., Richard Lupia, Andrew N. Drinnan, and Mark A. Burgman. 2001. The currency and tempo of extinction. *American Naturalist* 157, no. 1.

Reiger, John F. 2001. *American sportsmen and the origins of conservation.* Corvallis: Oregon State University Press.

Rolston, Holmes. 1994. *Conserving natural value.* New York: Columbia University Press.

Stegner, Wallace Earle. 1960 [1998]. "Wilderness letter" to David Pesonen. In *Marking the sparrow's fall: Wallace Stegner's American West,* ed. Page Stegner, 111–117. New York: H. Holt.

Takacs, David. 1996. *The idea of biodiversity: Philosophies of paradise.* Baltimore: Johns Hopkins University Press.

Tobin, Richard J. 1990. *The expendable future: U.S. politics and the protection of biological diversity.* Durham, NC: Duke University Press.

UNDP. n.d. *What is biodiversity?* New York: United Nations Development Programme. http://www.undp.org/biodiversity/biodiversitycd/whatIs.htm.

U.S.-Mexico Border XXI Program. 1996. *Framework document.* EPA 160-R-96-003. Washington, DC: U.S. Environmental Protection Agency.

Vogel, Joseph Henry. 1994. *Genes for sale.* New York: Oxford University Press.

Waller, Geoffrey, Michael Burchett, Marc Dando, and Richard Hull. 1996. *Sealife: A complete guide to the marine environment.* Washington, DC: Smithsonian Institution Press.

Ward, Peter. 1994. *The end of evolution.* New York: Bantam Books.

Weisman, Alan, and Jay Dusard. 1986. *La frontera: The United States border with Mexico.* San Diego, CA: Harcourt Brace Jovanovich.

Westing, Arthur H. 1993. Biodiversity and the challenge of national borders. *Environmental Conservation* 20, no. 1: 5–6.

Westing, Arthur H. 1998. Establishment and management of transfrontier reserves for conflict prevention and confidence building. *Environmental Conservation* 25, no. 2: 91–94.

Wilson, Edward O., ed. 1988. *Biodiversity.* Washington, DC: National Academy Press.

World Conservation Union. 2004. *The IUCN red list of threatened species: Summary statistics.* Gland, Switzerland: Author. http://www.redlist.org/info/tables/table1.html.

2

THE MULTIPLE PERSONALITIES OF TRANSBORDER CONSERVATION

North America's two major borders hold markedly divergent claims to fame. To the north, "the largest bilateral trading relationship in the world" extends across the Canada-U.S. border, which is also commonly described as the planet's longest undefended border (Canadian Embassy 2001). In contrast, the U.S.-Mexico border marks a discrepancy in per capita income greater than any other border on Earth. The reputations of both borders, however, result from their association with the world's last superpower. No doubt Mexicans can sympathize with Canadian Prime Minister Pierre Trudeau's famous quip that living next to the United States is "in some ways like sleeping with an elephant. No matter how friendly and even tempered is the beast . . . one is affected by every twitch and grunt" (Brunnée 2004, 617).

In addition to this similarity by association, the two borders share a characteristic that may surprise even experienced North American conservationists: Both are chock-full of efforts to conserve transborder biodiversity. Running your finger back and forth along a map of these two borders, you are touching on no less than thirty-seven such initiatives. Just look at the four termination points of the two borders: To the northwest, a massive World Heritage Area spans the Alaska–Yukon–British Columbia borders; the northeast has the Gulf of Maine Council for the Marine Environment; the southwest is home to the Tijuana River Watershed Project; and to the southeast lies the Laguna Madre Binational Initiative on the Gulf of Mexico. None of these places resounds with the cachet of the connotative domestic geographies of, say, British Columbia's Clayoquot Sound, Arizona's Grand Canyon, or the winter home of migratory monarch butterflies in El Rosario sanctuary of Michoacán's oyamel forests. And yet all are working toward the goal of biodiversity conservation and doing it in a transborder context.

Although the meaning of *transborder* may seem obvious, the word carries some semantic problems that are best addressed head-on. Throughout this book, I apply it to international relations involving border issues in a literal sense—water, air, and land issues that arise where one state ends and another begins. To understand exactly what distinguishes transborder issues, we could take an example from the world of international trade and finance: Whereas U.S. monetary policy for resolving the Mexican peso crisis is not a transborder issue, the efficient shipment of goods from Mexico to the United States is. Similarly, greenhouse gas emissions that lead to global climate change are not a transborder problem, but mercury emissions from an industrial facility that directly affect its downwind neighboring state are. In light of these examples, the term *transborder conservation* can be defined as the subset of international relations revolving around particular geographical borders that constitute some form of impediment to conservation. However, my definition of transborder conservation is neither rigid nor commonly held for at least two reasons: alternative labels and inherent ambiguity.

Transborder is one of numerous labels that scholars use to parse out the various forms of international relations. These include *transboundary*, *cross-border*, *transfrontier*, *transnational*, *transsovereign*, *interstate*, *intergovernmental*, and *transgovernmental*. Scholars apply different definitions to each of these labels in the literature on international relations; while only some of them take hold, what is perhaps most notable is the academic community's penchant for concocting labels—a good indication that international relations constitutes a far more complex beast than the stereotypical image of worldly diplomats staring each other down across a mahogany table. In any case, *transboundary* appears to have become more preponderant than *transborder* in the conservation literature, and two commonly used alternatives to *transborder conservation* are *transboundary natural resource management* (e.g., Griffin et al. 1999) and *transboundary ecosystem management* (e.g., Benvenisti 2002). The difference between these three is negligible, and the only reason I have chosen to go against the grain is because some have begun to use *transboundary* more broadly to describe interactions between subnational units—for example, between Australian states or Canadian provinces (see Griffin et al. 1999; for a thorough treatment of the meaning of *transboundary*, see Fall 2005).

With transborder conservation showing only inchoate signs of coalescing into a coherent field of study, there is no simple resolution to the problem of alternative labels. So although I generically rely on the term *transborder*, when citing others I will appropriate their language. And since this chapter provides a panoramic snapshot of what constitutes transborder conservation, such cases will be plentiful. Despite some terminological redundancy, this approach is more accurate than imposing some sort of linguistic order on the myriad transborder activities taking place across the globe.

Although my proposed definition of *transborder* may seem straightforward, its inherent ambiguity becomes apparent simply by juxtaposing it to the three major multilateral biodiversity conventions. First, the 1973 Convention on International Trade in Endangered Species (CITES) imposes trade restrictions without reference to particular geographic borders —yet much of the enforcement of CITES occurs at border checkpoints. So is CITES a transborder convention? Second, the 1979 Convention on the Conservation of Migratory Species of Wild Animals (known as the Bonn convention or CMS) certainly applies to species that cross borders, but the vast majority of work conducted under the Bonn convention focuses on protecting in-country habitats that are imperiled by land-use change (and not by a border per se). Is the CMS then a transborder convention? Third, the text of the 1992 Convention on Biological Diversity (CBD) never refers to transborder issues, yet the promotion of transboundary protected areas is now included in its "Programme of Work on Protected Areas" (Mulongoy and Chape 2004; Sandwith and Besançon 2005). Is the CBD a transborder convention?

To each of these questions, the most definitive answer I could offer is "Kind of." Although I do not focus on these global conventions in this book, each no doubt has a transborder aspect to it. The same holds true in regard to several higher-level international institutions—including the Organization for Economic Cooperation and Development, the European Community, and the UN Economic Commission for Europe—all of which by the late 1970s had at least proposed some form of transborder conservation activity (Wilcher 1980). Ultimately, it should come as no surprise that the topic has been adopted at such a broad scale; after all, at the heart of the subject lie roughly 220,000 kilometers (136,400 miles)—or five and a half Earth equators' worth—of international borders, many resounding with the strife of war, most with the mistrust of centuries, yet all with the hope of peace (Westing 1998, 91).

In sum, the proposed definition of *transborder conservation* establishes a somewhat porous circumscription around a certain type of international relations. Another way of getting at transborder conservation is to ask what differentiates it from domestic conservation. One might simply—and accurately—respond that the former is more complex than the latter, but a more thought-out response comes from Harris et al. (2001, iv). In examining eight case studies of "transboundary resource management efforts" in North America, they found "four main types of barriers that appeared to be much more prominent in transboundary situations": (1) legal and governmental differences that complicate coordination and implementation; (2) barriers to communication, movement, and information; (3) social and cultural differences, including language differences, that inhibit the development of trust and a common sense of community; and (4) economic disparities that constrain certain stakeholders' willingness or ability to participate in the process.

This list raises intimidating prospects and may at least partly explain why the entire issue of transborder conservation remains a minor sideshow within the off-Broadway theater of international environmental politics. As the remainder of this chapter documents, however, a growing number of conservationists are working on transborder protected areas, transborder biosphere reserves, and transborder bioregions. With biologists continuing to improve our understanding of the threats to landscape-scale biodiversity and the corresponding importance of landscape-scale conservation, it is a trend that seems likely to continue.

A Snapshot of Transborder Conservation

This chapter tells an abbreviated story of how conservationists have expanded their attention from relatively small transboundary protected areas (TBPAs) to larger—sometimes much larger—transborder landscapes and bioregions. First, however, it is important to point out that the universe of transborder conservation encompasses much more than TBPAs and transborder bioregions. This book's focus on the medium of land (that is, terrestrial biodiversity habitat) should not obfuscate the tremendous amount of international collaboration over the media of water and air.

Indeed, it is probably fair to say that the longest-standing, the most well-recognized, and the most contentious transborder issue is water. Fierce disputes over the quantity and quality of transborder water flows lie at the root of the burgeoning field of "environmental security" and "environmental peacekeeping" (Homer-Dixon 1999; Conca and Dabelko 2002). To resolve these disputes, diplomats and scholars have developed a convoluted web of international legal principles around the subject of *international watercourses*—which is the phrase that scholars use to signify the various forms of transborder water resources that include rivers, lakes, and even groundwater (Benvenisti 2002; Birnie and Boyle 1992, Chapter 6). Today, across a minimum of 261 international river basins worldwide (Wolf et al. 1999), approximately 300 international treaties "have been adopted for the purpose of avoiding conflicts over water" (Milich and Varady 1998, 10). Only a few of those treaties have attempted to go beyond the settlement of disputes over water rights toward the protection of either water quality or aquatic biodiversity.

Whereas transborder water issues are relatively delimited in terms of border geography, migratory species range over vast territories and multiple borders. As noted above, the CMS (as well as its subtreaties on migratory European bats and Eurasian waterbirds) essentially calls for habitat protection within member countries. That is, from a practical point of view, the CMS is largely about protecting point A and point B. But from a more optimistic vantage point, the treaties are also about protecting the very act of getting from point A to point B—the importance of which is hardly a recent discovery, since migratory species were critical in ushering in what one historian has described as the "dawn of conservation diplomacy" (Dorsey 1998). A glance at the history will help illuminate some of the roots of transborder conservation.

European efforts to protect an international flyway began as early as 1868, when an assembly of German farmers appealed to the Austrian and Hungarian foreign minister to join "other states in concluding an international agreement" for the protection of agriculturally beneficial birds that were being exterminated on one side of a border (particularly in Italy, a critical flyway to and from the Mediterranean), rendering protection on the other side impotent (Herman 1907, 32). Although the farmers' hopes were eventually realized in the 1902 International Convention for the Protection of Birds Useful to Agriculture (entering into force in 1905), Italy never signed on, and the treaty was generally considered ineffective (Heijnsbergen 1997, 10–11; Lyster 1985, 64).

Over in the Western Hemisphere, meanwhile, Canada and the United States had been wrangling for decades over the northern fur seal (*Callorhinus ursinus*), a migratory fur-bearing marine mammal that would come to hold a prominent position in the annals of conservation history. While many are today aware that President Theodore Roosevelt established Florida's Pelican Island as the nation's first national wildlife refuge in 1903, it is not widely recognized that thirty-five years earlier another Republican war hero, President Ulysses S. Grant, had established a "federal reservation" in order to protect the breeding grounds of the northern fur seal (Henderson 1901, 12; Merchant 2002, 228). Located on the Pribilof Islands in the Bering Sea (north of the Aleutian Islands), the breeding grounds had been included with the 1867 purchase of Alaska from Russia. Now part of the Alaska Maritime National Wildlife Refuge, the protected sections of the Pribilof Islands arguably beat out Pelican Island as the longest-standing component of the United States's National Wildlife Refuge system. Whatever the case, they do likely constitute the country's first occurrence of federal land protection for an "endangered species"—although that certainly was not the language used at the time (USFWS 1991, n.d.).

Although Grant's decision to protect the Pribilofs was reportedly confirmed by Congress in 1869, the protection did nothing to stop the decimation of pelagic sealing, and the issue soon became internationally contentious—even prompting the United States to seize Canadian schooners in 1886. A resolution to this crisis did not come until after the formative 1909 Boundary Waters Treaty between the United States and Canada (actually, Great Britain acting for Canada), which focused on navigation and water flows and established the International Joint Commission (IJC). Although this latter treaty did not initially address biological resources, it did set in place the main cornerstone of the relatively amicable relationship between Canada and the United States, thereby setting the standard for cooperation across North America (Spencer et al. 1981). Against this backdrop, the two countries signed the 1911 Convention for the Preservation and Protection of Fur Seals (with Russia and Japan as cosignatories). Although this convention was not "transborder" as defined here (since the seals did not cross a border from one country into another but rather moved from the northern Pacific Ocean onto U.S. territory), its success in turn set the stage for the 1916 Migratory Bird Treaty. Calling for an end to the hunting of certain taxonomic families of bird species, the treaty was ratified by Canada in 1917 and by the United States the following year. It is still in force today, is still causing controversy, and is considered by many conservationists a critical tool in bird conservation (Lee 2004, Williams 1997).

In 1936, the 1916 Migratory Bird Treaty was effectively extended to Mexico under the U.S.-Mexico Convention for the Protection of Migratory Birds and Game Mammals. Similar in its requirements to the 1916 treaty (notably, in 1913 U.S. officials had originally sought a treaty with Mexico before turning to Canada), the 1936 treaty is credited with being an essential stepping stone to later collaborative efforts between the United States and Mexico (Dorsey 1998, 196; USFWS and Direction General for Conservation and Ecological Use of Natural Resources of Mexico n.d.). As noted at the beginning of this chapter, such transborder collaborative initiatives have since become multitudinous, and it is difficult enough to keep track of all the ongoing transborder conservation efforts in North America alone. Nonetheless, before close examination of ISDA and Y2Y, it is worth adopting a global perspective on transborder conservation—particularly in regard to peace parks, transboundary protected areas, international biosphere reserves, and transborder bioregions.

A Plethora of Peace Parks

To counter the inevitable conflicts along the world's international borders, diplomats and conservationists have long pushed the idea of international "peace parks." The concept seems to have been initiated in Europe and not, as commonly thought, in North America. As early as 1780, a Treaty of Alliance between the King of France and the Prince-Bishop of Basel had stated that nothing "is more proper for maintaining good relations and peace between two bordering states" than punishing offenses related to forests, hunting, and fishing (Rüster and Simma 1975, 1541 [translation of Article 8]). Designating "an equal and uniform jurisprudence" over these issues within their shared border region, this treaty was also notable for stipulating that the two parties adopt the French Forest Ordinance of 1669, a law that had "sought to present a plan for sound land management to be applied to all of France" (Dasmann 1973–74, 473).

The modern concept of a peace park apparently originated in the 1924 Krakow Protocol between Poland and Czechoslovakia, an effort "to solve a boundary dispute that was one of the loose ends of the First World War"

(Westing 1998, 91). However, it took eight years for the protocol's call for international "reservation of regions for culture, wildlife, plant and local scenery protection" to bear fruit (Goetel 1964, 289). In the meantime on the other side of the Atlantic, a similar idea had reportedly occurred to park rangers in Glacier National Park (U.S.) and the adjacent Waterton National Park (Canada)—as well as to individual members of the Cardston Rotary Club in Canada (MacDonald 2000; Rotary International et al. n.d.; Stewart n.d.).

About thirty miles east of Waterton, Cardston was a small Mormon town in southern Alberta, and its Rotary Club was one of hundreds scattered around the United States and Canada. At a goodwill meeting in Waterton on July 4, 1931, Rotary Clubs from Alberta and Montana formally proposed an international peace park between Glacier and Waterton. Rotarians on both sides of the border immediately turned to lobbying their respective governments. The official response was remarkably quick, as both the Canadian Parliament and the U.S. Congress were able to pass legislation in time for a dedication ceremony in Glacier Park on June 19, 1932 (Rotary International et al. n.d.; Stewart n.d.). Since then, an annual assembly of local Rotary Clubs has alternated between the two parks, each meeting culminating in a "hands across the border" ceremony (Stewart n.d.).

Two months after the dedication of Waterton-Glacier, Poland and Czechoslovakia formally recognized "the first International Landscape Park in Europe" between the Polish Pieniny National Park and the Slovak National Natural Reserve on August 17 (Kukura n.d.; Sochaczewski 1999, 36). With the establishment of the world's first two international peace parks, 1932 can be described as the watershed year for peace parks.

Only a year later, other European colonial powers were considering the idea of transborder conservation—but interestingly enough, not in Europe. In 1933, the colonial powers signed the London Convention Relative to the Preservation of Fauna and Flora in their Natural State (Heijnsbergen 1997, 13). Entering into force in 1936, this convention was a follow-up effort to what appears to have been the first multilateral convention on international conservation: the 1900 Convention for the Preservation of Wild Animals, Birds and Fish in Africa (*Times of London* 1900). As "the first treaty to encourage the establishment of nature reserves" and the first to list which species were to be protected, the 1900 convention substantively prefigures the CBD and CITES (Lyster 1985, 113). Yet, forward thinking as it was, the 1900 convention had no specific transborder clauses (and in any case, it never entered into force). In contrast, the 1933 convention called for "prior consultation" and "cooperation" where "a national park or strict natural reserve" had been or was to be established "contiguous to a park or reserve situated in another territory . . . or to the boundary of such territory" (Article 6). Although the colonial powers probably had certain areas in mind in drafting such specific language, I have yet to encounter any direct or indirect reference to its implementation.

In 1968, the newly independent African countries replaced the London convention with the African Convention on the Conservation of Nature and Natural Resources, which does not refer directly to transborder conservation—an unfortunate omission, given that sub-Saharan Africa would subsequently become a global focal point for the establishment of several large peace parks. While each of those parks deserves attention, no history of African peace parks would be complete without at least briefly touching on the saga of the Virunga Mountains region. In 1925, Belgium established Africa's first national park, which originally covered the western half of the Virunga Mountains in the Belgian Congo. The park was named after the ruling King Albert, who, having been inspired by several prominent conservationists on a visit to Yellowstone National Park in 1919, had been subsequently convinced by American naturalist Carl Akeley to protect the central African area for its mountain gorillas (Sanwekwe, a Congolese boy who helped Akeley track gorillas, would go on to work with two other noted American gorilla biologists: George Schaller and Diane Fossey) (Fitter and Scott 1978, 16–17; Fossey 1983, 2; Weber and Vedder 2001, 42). In 1929, the Belgian authorities expanded Albert National Park to include all of the Virunga Mountains traversing the two colonies of the Belgian Congo and Ruanda-Urundi (a League of Nations territory under Belgian rule). This expansion laid the foundation for an incipient transborder park, for when the colonies gained their independence in the 1960s, the park was split into the Virunga National Park of the Democratic Republic of the Congo (Zaire 1971–1997) and Volcanos National Park of Rwanda. Although transborder conservation initiatives in the region have been stymied by civil wars on both sides of the border, in October 2005 the two countries along with Uganda signed a Tripartite Declaration that recognized the need to establish a "Central Albertine Rift Transfrontier Protected Area Network" (AWF 2005; Besançon and Hammill, forthcoming; Lanjouw et al. 2001; Wilkie et al. 2001).

Peace Parks as the Public Face of Transborder Conservation

Notably, while Article 6 of the 1933 London convention concerned the coordinated management of parks and reserves in Africa, it did not refer to the subject of peace. This raises an important point: Although "peace" and "park" make mutually admirable goals, they are not one and the same thing. At a minimum, as Gartlan (1999, 247) has argued, the "emergent" function of promoting peace is "*additional* to the functions of conservation and habitat protection" (adding that the peace function "is essentially an institutional one to be negotiated and agreed upon between the highest levels of government . . . at a later stage"). Yet for several reasons, *peace park* will no doubt remain the most common term for site-specific transborder collaboration. For one thing, the phrase is alliterative and colorful. Furthermore, depending on the level of strife involved in any particular site, conservationists will recognize that the "peace" side of the equation will normally achieve greater visibility than the "park" side (i.e., matters of biological conservation). Few conservationists will hesitate to ride the coattails of what will almost always prove to be the issue of paramount political significance. Finally, early synthesizing scholarship on transborder conservation tended to focus on peace parks, particularly the work of Arthur Westing. After leading a broad investigation into the relationship between war and the environment in the wake of the seminal 1972 Stockholm conference, Westing became a recognized authority and strong proponent of peace parks (Hayes 2001; see citations in Westing 1998).

Despite the predominance of the label, *peace park* is only one name among many. Others include *border park*, *transborder park*, *borderline park*, *friendship park*, *transnational park*, *transfrontier reserve*, *transboundary conservation area*, *transborder conservation area*, *cross-border protected natural park*, and *transboundary natural resource management area* (Griffin et al. 1999; Tolentino 2001, 43; Westing 1993b). Out of this titular thicket, the generic *transboundary protected area* has become the most widely accepted term in policy and scholarly circles. Again, I largely follow this convention when directly citing others.

The IUCN makes a point of distinguishing between peace parks and TBPAs (Sandwith et al. 2001, 3). A peace park is defined as an area "formally dedicated to the protection and maintenance of biological diversity, and of natural and associated cultural resources, and to the promotion of peace and co-operation." The definition of a TBPA, in contrast, manages to be both more delineated and more expansive at the same time:

An area of land and/or sea that straddles one or more boundaries between states, sub-national units such as provinces and regions, autonomous areas and/or areas beyond the limits of national sovereignty or jurisdiction, whose constituent parts are especially dedicated to the protection and maintenance of biological diversity, and of natural and associated cultural resources, and managed co-operatively through legal or other effective means. (Sandwith et al. 2001, 3)

Despite its delineation, this is a fairly open-ended definition. One could easily set the bar much higher by defining TBPAs as entailing consistent management plans, joint working-level consultative committees, harmonized law enforcement, or "a sustainable financing strategy" (Tolentino 2001, 56). Alternatively, one could simply give up on defining TBPAs on the grounds that "at least from a legal and administrative point of view, it is more accurate to speak of transboundary cooperation in protected areas rather than transboundary protected areas" (Brunner 1999). But overall, the IUCN's balanced definition best captures the commonly accepted meaning of the term.

How many TBPAs are there in the world? Trying to count them all has proven challenging. During the 1980s, the IUCN identified approximately 70 protected areas sharing "common international boundaries," which translated into a total of approximately 35 transborder areas (Douglas 1997, 31). In 1990, it was estimated that there were "50 pairs or so" of "transfrontier nature reserves" (Thorsell and Harrison 1990). By the mid-1990s, "at least 100" TBPAs had been identified (Hamilton et al. 1996, 1). An extensive search and analysis commenced in 1997 by Dorothy Zbicz, now a consultant to the World Wildlife Fund, ultimately resulted in a 2003 listing of 169 transborder protected area "complexes" involving 666 individual protected areas and 113 different countries and covering "at least 10 percent of the total area of all the protected areas in the world" (United Nations Environment Programme et al. 2004; Zbicz 2003, 24). Finally, in 2005, Besançon and Savy listed 188 "internationally adjoining protected areas." Note that although the growth in TBPAs was real, it was not as rapid as these figures might indicate, since the earlier assessments likely undercounted the number worldwide.

With so many TBPAs worldwide, conservationists have devised several ways of sorting them out. One group of conservation practitioners, for instance, has delineated five different types of TBPAs on the basis of geographic parameters:

- two or more contiguous protected areas across a national boundary;
- a cluster of protected areas and the intervening land;
- a cluster of separated protected areas without intervening land;
- a transborder area including proposed protected areas; and
- a protected area in one country aided by sympathetic land use over the border (Bakarr 2003).

Alternatively, Tolentino (2001) divided TBPAs into three developmental types: the first established through legal instruments (such as treaties), the second through institutional arrangements (such as commissions), and the third through both. Sandwith et al. (2001, 7–12) also point to three different ways that "transboundary initiatives develop." First, *high-level initiatives* involve officials within an administrative capacity above the level of direct land management. Second, *locally based initiatives* refer to those established at the level of direct "on-the-ground" land management (the authors note, for example, that "transboundary co-operation may begin with two individual field staff members who experience real benefit through co-operation in one or more specific tasks on the ground, such as fire suppression"). Finally, *third-party initiatives* occur "*via* a conservation non-governmental organisation (NGO) acting as a third party advocate, encouraging and supporting co-operative transboundary management."

Hamilton et al. (1996, 13–15) provided a similar but more nuanced list and proposed an incipient typology based on level of cooperation between land managers and staff of internationally adjoining PAs (Zbicz 1999). Zbicz (2003, 25–26) then grounded that approach by identifying six "hierarchical, increasing levels of transboundary cooperation between adjoining protected areas": no cooperation, communication, consultation, collaboration, coordination of planning, and full cooperation. In order to apply this system to the aforementioned global database of TBPAs she had already developed, she conducted a global survey of managers working in TBPAs, asking for their perspectives on a wide range of transboundary activities and attitudes. Based on the responses to the survey, she placed each complex of PAs within one of the six levels of transboundary cooperation. At the extremes, 18 percent responded that there was no cooperation at all, while 7 percent were cooperating at the level of "full cooperation" (Zbicz 2003). The largest minority, 39 percent, was at the level of "communication."

In her analysis, Zbicz drew out four "factors" correlated to the level of cooperation. In essence, higher levels of cooperation occurred (1) if the idea of transfrontier cooperation and ecosystem-based management was important to the protected area managers and personnel, (2) if there were adequate communication technologies in place, (3) if there were individuals willing to take leadership roles, and (4) if land managers were able to make personal contact across the border. Not surprisingly, it was the latter factor that correlated most strongly with the level of cooperation achieved. "In transboundary cooperation," writes Zbicz, "face-to-face contact is not only the preferred means of communication, but also often the *only* available and acceptable means" (Zbicz 2003, 33). Notably, Zbicz also found that higher-income countries did not "appear to value transfrontier cooperation or ecosystem-based management much more than lower-income countries" (2003, 29).

Pros, Cons, and "Best Practices" of TBPAs

Practitioners and observers have drawn up a list of over twenty possible benefits of transborder cooperation (Box 2.1; Sandwith et al. 2001). Although wide ranging—it includes promotion of "ecosystem-based management" as well as sharing expenses for "infrequently used heavy equipment"—the list is not exhaustive. Louka (2002, 30), for instance, adds that "transnational management" can "become a way to introduce more checks and balances and more transparency in the management of resources. . . . Human rights violations in transnational parks are likely to gain more publicity, making transnational management a deterrent to the assaults on human dignity."

Despite these multiple benefits, TBPAs have been criticized on both social and biological grounds, and even their proponents have allowed that they face numerous "inevitable impediments" that include conflicting laws, language barriers, differing stages of economic development, discrepancies in discretionary authority, and uncertainty over "whether the costs involved in these complex programmes really justify the biodiversity, social, institutional, political and economic benefits" (Hamilton et al. 1996, 6–7; Transboundary Protected Areas Task Force 2004, 16). Several critiques on social grounds have come from Africa, where numerous

Box 2.1
Benefits of Transboundary Protected Area Cooperation

1. A larger contiguous area will better safeguard biodiversity, since very large areas are needed to maintain minimum viable populations of many fauna species, particularly large carnivores.
2. Where populations of flora or fauna cross a political or administrative boundary, transboundary cooperation promotes ecosystem or bioregional management.
3. Reintroduction or natural recolonization of large-range species is facilitated by transboundary cooperation.
4. Pest species (pathogens, insects) or alien invasives that adversely affect native biodiversity are more easily controlled if joint control is exercised rather than having a source of infection across the boundary.
5. For rare plant species needing ex situ bank and nursery facilities, one facility for both parks will be cheaper to set up than separate ones.
6. Joint research programs can eliminate duplication, enlarge perspectives and skill pools, standardize methodologies, and share expensive equipment.
7. Wildfires cross boundaries, and better surveillance and management are possible through joint management.
8. Poaching and illegal trade across boundaries are better controlled by transboundary cooperation, including joint patrols.
9. Transboundary cooperation enhances nature-based tourism because of a greater attraction for visitors, possibilities of joint approaches to marketing and tour operator training, and the possibility of agreements on fees, visitor management, etc.
10. More cost-effective and compelling educational materials can be produced, and joint interpretation of shared natural or cultural resources is stronger.
11. Joint training of park staff is more cost effective and usually benefits from greater diversity of staff with different experiences.
12. Transboundary cooperation improves staff morale and reduces the feeling of isolation. Contact with cultural differences enriches both partners.
13. Transboundary cooperation makes staff exchanges easier; staff exchange programs have shown their worth.
14. A cross-boundary pool of different types of expertise is available for problem solving.
15. Expenses for infrequently used heavy equipment, aircraft rental for patrols, etc., may be shared.
16. Transboundary cooperation in priority actions can carry more weight with authorities in each country.
17. The ministry level may feel a greater obligation to honor commitments of support when another jurisdiction or another country is involved.
18. International designation, donors, and assistance agencies are more attracted to an international joint proposal.
19. Outside threats (e.g., air pollution, inappropriate development) may be more easily met by an international or interstate response.
20. Customs and immigration officials are more easily encouraged to cooperate if parks are cooperating.
21. Search and rescue efforts are often more efficient and economical.

Source: From Sandwith et al. (2001, 8), based on Hamilton et al. (1996).

proposed and actual TBPAs intersect with extreme poverty and significant human rights violations. The critiques range from a general accusation that peace parks represent little more than a contemporary colonialist attitude toward Africa (e.g., Draper et al. 2004) to the more specific argument that they have the ironic effect of actually fostering animosity (e.g., disputes over revenue sharing from ecotourism) (Kahn 2003). Dutton and Archer (2004) have also argued that despite the progress made in social equality in South Africa, the movement toward transfrontier parks has not taken into account the rights of the country's indigenous people. Rogers (2002) has pointed to multiple other problems arising from the imposition of an external conservation agenda on indigenous peoples in east and southern Africa.

Although the evidence is scanty, TBPAs have also been criticized on biological grounds. In one of the few extant biological assessments, Reyers (2003) found that TBPAs along South Africa's borders did not produce significant benefits either in increasing vegetational representation or in protecting additional bird species. While she noted that her data covered only one aspect of biodiversity conservation, she argued that "TBPAs could actually undermine the objectives of regional biodiversity conservation and could in effect amount to ad hoc land allocation" that garners little actual benefit. More generally, several conservation practitioners have voiced concern that, as Gary Tabor of the Wilburforce Foundation has put it, TBPAs will unduly shift conservation priorities on the basis of political imperatives rather than conservation imperatives. That is, they are concerned that a focus on TBPAs could divert resources away from particular domestic conservation activities that ultimately have a higher conservation value.

TBPAs have also become a flashpoint in a larger debate within the conservation community over the relationship between conserving biodiversity and meeting the needs of an ever-growing human population. Running congruent to endless deliberations over what exactly constitutes "sustainable development" (such discussions being much older than the label itself), the debate can be simplistically divided into two ideologies. On one side are the challengers, who argue that the traditional form of conservation—that is, putting a wall around "nature" and excluding all but the most transient of human visitors—can work only in limited conditions within the developed world. Rather than throwing barricades around biodiversity, they argue, long-term protection depends on ensuring that people can live sustainably off the habitat to be protected, or at least can find gainful livelihoods in some other fashion. Practical implementation of this approach has come under the banner of "community-based conservation" (CBC) and "integrated conservation and development projects" (ICDPs), the latter of which is generally defined as biodiversity conservation projects with rural development components that are located near protected areas (Hughes and Flintan 2001, 4). The opposite encampment, the defenders, has responded that the myriad attempts at CBCs and ICDPs have generally failed to protect biodiversity and that protecting all the components of biodiversity in any given ecosystem requires direct habitat protection. Furthermore, they argue, traditional efforts to protect habitat already provide many direct and indirect benefits to local people.

The literature spawned by this debate is copious and, of course, far more nuanced than the above summary suggests. TBPAs have become embroiled in the debate under the accusation that they emanate from the defenders' encampment and thus constitute little more than a new approach to bottling up nature. Brosius and Russell (2003, 40), for instance, believe that the transborder approach "has resulted in part out of a loss of faith in the idea of community-based conservation, a faith that may never have had many true believers in large conservation organizations in the first place." Practitioners of transborder conservation disagree vehemently, arguing that human betterment has always been an integral component of their efforts to establish TBPAs.

To address the practical and ethical problems posed by TBPAs, numerous sets of formal and informal "best practices" have been developed. As early as 1980, the Council of Europe agreed to a *European Outline Convention on Transfrontier Co-operation between Territorial Communities or Authorities*, Section 1.9 of which consisted of a "Model Agreement on the Creation and Management of Transfrontier Parks." The Model Agreement called for the parties "to harmonise their methods of management and to co-ordinate all development projects or improvements by means of a comprehensive action programme leading ultimately to joint management of the park based on a joint management plan." The Model Agreement also called for joint committees whose membership would include "representatives of recognised private nature conservation organisations and organisations which contribute to the safeguarding of the landscape and the environment" (Council of Europe 1980).

From a broader perspective, Westing (1998, 93) has proposed both "environmental and political criteria" for transfrontier reserves that "are equally consequential, clearly neither alone sufficing." On the environmental side, potential sites should be assessed on one or more of three factors: whether the area has been identified as a "hot spot" for biodiversity, whether it is located in an inadequately protected biome, and whether at least one of the countries involved has an "inadequate proportion of its territory in a protected state." In addition to these environmental criteria, three political criteria bear on the viability of a potential peace park: "whether it would safeguard and improve existing friendly relations between two (or more) neighbouring states; whether it would make irrelevant an existing dispute over contested land or territorial waters; or whether it would help bring divided states back together." Westing further proposed that transfrontier protected areas should be established where "population pressures on both sides are roughly similar," where there is "comparability of way of life between the two sides," and where protected areas on both sides of a border "have a similar conservation status" (see also Gartlan 1999, 247).

A more explicit set of general guidelines for transboundary cooperation was set forth in 1988 at a workshop on transboundary areas (Brunner 1999, 28). By the mid-1990s, conservationist Lawrence Hamilton of the IUCN's World Commission on Protected Areas (WCPA) had begun promoting "best practice guidelines" for transboundary cooperation (see citations in Hamilton 2001). Most recently, the WCPA put forth a set of "good practice guidelines" under nine primary headings:

- identifying and promoting common values;
- involving and benefiting local people;
- obtaining and maintaining support of decision makers;
- promoting coordinated and cooperative activities;
- achieving coordinated planning and protected area development;
- developing cooperative agreements;
- working toward funding sustainability;
- monitoring and assessing progress; and
- dealing with tension or armed conflict (Sandwith et al. 2001, 17).

Along with these guidelines, the WCPA proposed a "Draft Code for transboundary protected areas in times of peace and armed conflict," which essentially constitutes an annotated template for a formal bilateral agreement over a transboundary protected area (Sandwith et al. 2001, 39–52).

In 2003 the EUROPARC Federation (formally known as the Federation of Nature and National Parks of Europe) established a certification system for "exemplary transboundary cooperation between protected areas" according to a set of criteria in the form of seven questions:

- Do the parks have a common vision for sustainable development in the region?
- Is an agreement in place, which is signed by the parks or at political decision-making levels and which guarantees the continuity of the cooperation?
- Does a joint work program exist, which defines the main areas of cooperation in the individual fields of work?
- Are mechanisms for direct cooperation between protected area staff, the regular exchange of experience, and the implementation of joint meetings and decisions established?
- Does observation of changes in parks' natural values through joint monitoring and the holding of regular exchanges of data take place?
- Are steps taken to ensure that communication between the protected areas is not held back by language barriers?
- Are joint transboundary projects in existence and has their financing been secured?

EUROPARC has argued that these criteria have already been useful in "assessing the state of cooperation between neighbouring protected areas" and that "the system can therefore already be regarded as an important motor for the further development of transboundary cooperation and can be applied—with some modifications—in other continents too" (EUROPARC Federation 2003).

International Support for TBPAs

At least one observer has called for an international "legal regime" on transborder parks (Tolentino 2001). Westing (1993a; 1993b, 4–7) has found some legal foundation for such an idea in four multilateral and numerous regional and bilateral agreements that in some way concern "the explicit protection of transfrontier natural habitats" (see also Tolentino 2001, which lists several additional agreements). Although the text of the Convention on Biological Diversity does not refer to TBPAs, in 2004 the countries that have ratified the convention (known as the "Conference of the Parties") adopted the goal of establishing and strengthening "regional networks, transboundary protected areas and collaboration between neighbouring protected areas across national boundaries" under its "Protected Areas Programme of Work" (Convention on Biological Diversity 2004, 353). In addition, the IUCN's 2004 draft of an International Covenant on Environment and Development states that parties to the convention would "cooperate in the conservation, management and restoration of natural resources which occur in areas under the jurisdiction of more than one State, or fully or partly in areas beyond the limits of national jurisdiction. To this end, (a) Parties sharing the same natural system shall make every effort to manage that system as a single ecological unit notwithstanding national boundaries . . ." (Commission on Environmental Law 2004, Article 35).

Overall, although there are some precedents for a legal regime on TBPAs, such a framework is unlikely to emerge out of the already crowded arena of international environmental law. Nevertheless, institutional support for TBPAs has arisen from within the international community. Two major international financial institutions, for example, have focused their efforts on TBPAs. The World Bank has financially supported a number of transborder protected area projects and investigations (Mackinnon 2002), and the International Tropical Timber Organization (ITTO) has funded a minimum of seven transboundary conservation areas (International Tropical Timber Organization 2004; Poore 2003, 236–238).

Finally, and perhaps most important, a community of dedicated researchers and practitioners has materialized around TBPAs, holding at least seven significant conferences and meetings focused on the subject since 1988. This community has manifested itself in several interrelated institutional bodies. First, in 1997 a Parks for Peace initiative was established as a joint undertaking of the South African Peace Parks Foundation and three arms of the IUCN (the WCPA, the Programme on Protected Areas, and the Commission on Environmental Law) (Sandwith et al. 2001, 1). From that collaboration arose the WCPA's Task Force on Transboundary Protected Areas, the goals of which include creating new peace parks, strengthening existing ones, establishing best practice guidelines and a code of conduct for peace parks, and establishing a permanent Peace Parks Council and a Global Partnership for Peace Parks that would build on the experience of the IUCN, World Wildlife Fund–International, the Costa Rica–based University for Peace, and the Peace Parks Foundation (World Commission on Protected Areas 2003). In addition, the Task Force is currently evaluating the possibility of the IUCN adopting transboundary protected areas as a major thematic area (Transboundary Protected Areas Task Force 2004). Along with the Global Environment Facility (GEF), the International Tropical Timber Organization, and several European state agencies, the Task Force has organized a Global Transboundary Protected Area Network (Global Transboundary Protected Areas Network n.d.). The network's Web site currently constitutes the most comprehensive collection of TBPA information available.

Bioregions and the Biosphere

Due to both the geographical and the conceptual restrictiveness of TBPAs, many conservationists are refocusing their efforts on transborder regions—regions much larger than those traditionally considered under the peace park aegis. For example, in arguing that the issues of conservation and human rights are inextricably intertwined, Louka (2002, 30) proposed that the term *transnational protected area* be replaced with *transnational landscapes*. Similarly, in a project involving Mozambique, South Africa, and Zimbabwe, the Global Environment Facility expanded its focus from "the original idea of a transborder park" into "a broader multiple use 'Transfrontier Conservation Area' (TFCA) concept." The GEF then defined TFCAs as "relatively large areas, which straddle frontiers between two or more countries and cover large scale natural systems encompassing one or more protected areas. These are areas where human and animal populations have traditionally migrated across or straddled political boundaries" (Global Environment Facility 1996, 5–6).

In refocusing their efforts on larger regions, both Louka and the GEF were in essence calling for a form of bioregionalism. According to Kirkpatrick Sale (1985), an early advocate of bioregionalism, the idea's roots run deep to prehistorical societies. Recounting the evolution of the idea up to its role in modern American culture, Sale notes that the basic idea of "regionalism" has become generally associated with historical theories of the country's development (e.g., Frederick Jackson Turner's sectionalism theory) and with efforts to instill "regional planning" as a way of curbing the nation's ills (e.g., Lewis Mumford's Regional Plan

Association of America). Sale argues that despite waxing and waning over the course of the century, regionalism has been a constant undercurrent in both American culture and global affairs.

Moving from this sociological perspective, Sale then poetically describes bioregionalism as about people becoming "dwellers in the land" who "come to know the earth fully and honestly" (Sale 1985, 42). But he also points out that bioregionalism finds much of its intellectual heritage in the natural sciences, and indeed *bioregionalism* and its cousin *ecoregionalism* have become common terms in the fields of ecology and conservation (Bailey 2002). As Franklin (1999, xii; see also other chapters in this edited volume) has written, these ideas have been adopted by scientists who had been "catapulted into central roles in natural resource policy analysis by the need—legal and social—for development of scientifically credible plans for management of wildland ecosystems over entire regions."

Sale attributes the relatively recent conceptualization and use of the term *bioregion* to the collaboration between writer Peter Berg and ecologist Raymond Dasmann during the 1970s—a collaboration that led to the establishment of Planet Drum, an organization that has promoted bioregional initiatives around the world (Sale 1985, 43). The involvement of Dasmann, an author of several books on conservation (including a 1959 textbook entitled *Environmental Conservation*) and a well-respected educator (one of his students would go on to head ISDA), is particularly notable in that it points to a significant scientific thread in the history of bioregionalism (Jarrell 2000). That particular thread, in turn, is directly tied to the conceptual origins of this book's case studies of the International Sonoran Desert Alliance and the Yellowstone to Yukon Conservation Initiative. For previous to helping coin "bioregion," Dasmann had played an active role in the development of the Man and the Biosphere Program (MAB) and the paradigmatic concepts it set forth—concepts that were fundamentally bioregional and that would heavily influence the development of both ISDA and Y2Y. Importantly, however, ISDA and Y2Y would adopt the core MAB ideas in very different ways from each other. To understand these differences, it is necessary to take a closer look at the origins of the MAB Program.

Man and the Biosphere

The origins of the MAB Program are typically traced to UNESCO's 1968 International Conference of Experts on a Scientific Basis for Rational Use and Conservation of the Biosphere (Batisse 1997; de Klemm and Shine

1993), but its intellectual roots run much deeper than the "Biosphere Conference." They go as far back as the 1882 International Polar Year (IPY), an international scientific collaboration between eleven countries that established an incipient network of arctic research stations (Stone and Vogel 2004). The project was repeated in 1932–33, and a proposed third IPY for 1957–58 was renamed the International Geophysical Year (IGY), which blossomed into a global program of geophysical research across fourteen disciplines that involved sixty-seven countries with over 2,500 stations worldwide (Berkman 2003).

These successful scientific collaborations inspired biologists in 1959 to begin planning their own version of a large-scale international research effort: the International Biological Program (IBP). Conducted under the aegis of the International Council of Scientific Unions (and partially supported by the U.S. National Science Foundation), the IBP would extend over the course of a decade (1964–1974) (Worthington 1975). Between 1959 and 1964, a significant planning venue for the IBP was the IUCN, where science had been considered integral to the institution's conservation mission since its establishment in 1948 (Holdgate 1999, 32–33, 93–96). Yet despite the fact that one of the IBP's seven broad research "sections" was entitled *Conservation Terrestrial*, "the institutional links between IUCN and IBP remained slender" (Holdgate 1999, 95). The "Conservation Terrestrial" section apparently received the least enthusiastic reception from IBP scientists, and Dasmann would later describe IBP as "purely a scientific program" that "was fine as a research effort but not in the sense of changing anything on the ground" (Dasmann 2002, 140). Yet over the course of its decade-long life, IBP came to narrow its focus to the systematic study of ecosystems—a focus that not surprisingly would have significant implications for conservation (Golley 1993, 112).

Halfway through the course of the IBP, Dasmann entered the picture through a complex web of connections. In 1966, the UN Economic and Social Council (ECOSOC) had "invited" UNESCO to prepare a report that would eventually be titled *Conservation and Rational Use of the Environment*, a subject that UNESCO had been considering for a global conference. The invitation was directed to UNESCO's Division of Natural Resources Research, which was then led by engineer and physicist Michel Batisse, a pioneer in international environmental issues who would become a pivotal promoter of MAB and whose career at UNESCO would last a half century (UNESCO 2005b). Having been working closely with the IBP,

Batisse apparently recognized this as an opportunity to begin consideration of who or what would take over all of the IBP's work once it came to an end—even though at this point the IBP was not half through. Batisse then turned to two prominent ecologists, Gilbert White and Luna Leopold (Aldo Leopold's son), to recommend an author for the report (Holdgate 1999, 97). They recommended Dasmann, who at the time was taking "a leading role in relating the content of ecology to conservation policy" as a staff member of the Conservation Foundation (Dasmann 1966, 326). Batisse pleaded for help from Russell Train, director of the Conservation Foundation, and Dasmann was assigned to the job (Dasmann 2002, 139). Regardless of any concerns over IBP's practical efficacy that he may have harbored at the time, Dasmann recognized "the need to launch an intergovernmental program to carry on where the IBP would leave off" (Dasmann 2002, 140). Along with UNESCO and IUCN personnel, IBP scientists would play a significant role in running the 1968 Biosphere Conference in Paris, at which Dasmann's report served as a background document for the assembled participants (Adams 1990, 33; Dasmann 2002, 140).

Adams (1990, 30–36) describes this conference as a key moment in the coalescing of development goals with environmental goals—that is, as one of the major intellectual precursors of the idea of "sustainable development." The Biosphere Conference provided the impetus for the General Conference of UNESCO to charter the International Coordinating Council (ICC) for the Man and the Biosphere Program in 1970 (Holdgate 1999; UNESCO 2003). Launched the following year, MAB would come to be described as "the first deliberate international initiative to find ways to achieve sustainable development" (Gregg 1999, 24).

To some degree, MAB's conceptual basis was rooted in the perceived inadequacy of the IBP. In particular, Golley (1993, 163) describes MAB as addressing three criticisms of the IBP:

> The lack of practicality had been a criticism of IBP by the ecologists of the USSR and the Central European countries, and while IBP was intended to understand the basis of biological productivity for human welfare, it was quickly captured by ecologists of dominant western countries who created an academic program that explored the reigning ecological paradigms. MAB reversed this emphasis, building upon the substantial advances in knowledge that came from IBP research. Second, MAB focused on developing countries and

the ecosystems in those countries that had been neglected or underemphasized in the IBP. For example, MAB set research on tropical forests as its highest priority, with arid and semiarid grasslands and deserts as its second priority. Third, MAB studied systems in which humans were an integral part, including cities, agricultural systems, and natural reserves. . . .

Golley (1993, 162) goes on to explain that the administrative shift from IBP under the International Council of Scientific Unions to MAB under UNESCO was "extremely significant because it meant a shift from academic science and individuals to governmental science." As UNESCO's Batisse emphasized, although scientists would carry out MAB, the program would ultimately be sponsored by "governments in the service of man" (Harrison 1982).

Development of Biosphere Reserves

While the 1968 Biosphere Conference had recommended the setting aside of natural areas, it was only the following year that Batisse apparently first suggested the specific idea of a "biosphere reserve" (Fall and Andrian 2005). In 1971, the MAB ICC formally proposed an international network of biosphere reserves, a concept that was endorsed the following year at the Stockholm Conference on the Human Environment (Gregg 1999, 25). At a 1973 planning meeting in Morges, Switzerland, the ICC made the establishment of a worldwide biosphere reserve network its first recommendation. This was despite the fact that at its conception, the MAB Program had been divided into thirteen different "projects," and biosphere reserves were administratively placed within Project 8: Conservation of Natural Areas and of the Genetic Material They Contain (Franklin 1977; Gregg 1999, 24). In 1974, five countries nominated a total of 24 biosphere reserves—20 of them by (and within) the United States (UNESCO 1974a). Over thirty years later, with 482 biosphere reserves in 102 countries as of November 2005, the MAB Program has become nearly synonymous with biosphere reserves (UNESCO-MAB Secretariat 2005).

What did its early conceivers envision for the biosphere reserve network? The development of MAB's orientation toward biosphere reserves during those early years has been attributed to Francesco di Castri, a principal investigator and ecologist under the IBP who became instrumental in MAB's creation—as well as its first director (Golley 1993, 162). In my

2000 interview with Exequiel Ezcurra, a prominent Mexican biologist and high-ranking government official (and a prominent character in the story of ISDA; see Chapter 3), he described di Castri as a Renaissance man whose leadership during the 1970s constituted a "golden period" for MAB. Di Castri pushed the MAB Program to understand that "conservation had been done the wrong way—that conservation had been conceived in developed countries as very much following the paradigm or the model of a flask or a jar, that you could put nature in an isolated jar and it didn't really matter what happened outside, but this microcosm would survive. He said this just does not happen in the developing world because you cannot keep an isolated jar with the world around it going to pieces—because sooner or later somebody will break the jar and come in and take over resources." (Ezcurra added at this point: "Even the name that we still use a lot in the English language—preserve, environmental preserves—it sounds like a marmalade jar or something. Something that you seal very tightly and boil.") Instead of a series of microcosms, Ezcurra continued, di Castri's original intent for the MAB program was to protect the *global* biosphere by protecting the "largest biomes on Earth." Indeed, said Ezcurra:

> That's why they're called biosphere reserves—because they're a network of reserves that are aimed at protecting the biosphere. Which is funny—many people have lost the whole idea of the origin of the name. People will say, "oh, this is a 'biosphere'" because it's [designated as] a "biosphere reserve." So they think that that place is *a* biosphere. No, the idea of biosphere reserves was a series of reserves that jointly made a system that could protect the *whole* biosphere.

Although there is some documentary evidence for di Castri's global biosphere perspective (see Rikoon and Goedeke 2000, 24), di Castri's perspective never became widespread.

A set of "criteria and guidelines" for biosphere reserves were established at a task force meeting of the ICC in Paris in 1974 (Franklin 1977). At that meeting, three fundamental goals of biosphere reserves were put forth: (1) to contribute to the conservation of ecosystems, biotic communities, species, and genetic diversity; (2) to provide research areas for "baseline" ecological and environmental studies; and (3) "to provide facilities for education and training" (UNESCO 1974b, 11–12). Although it is not clear exactly when, by the 1980s the latter two had been combined under the all-purpose objective of "logistics," and a third objective of development—

now, of course, sustainable development—was articulated (Hough 1988, 1). These three goals have been reinterpreted in various forms, often with significant variations of the logistics objective.

It was in the attempt to implement these three goals that the "biosphere reserve ideal" influenced the development of both Y2Y and ISDA. As noted previously, however, that influence differed strongly between the two. Whereas the basic conceptualization of land management patterns within biosphere reserves shaped much of the thinking behind Y2Y, ISDA was influenced more by the emphasis on local participation that MAB adopted in its second decade.

Biosphere Reserves, Land-Use Patterns, and Y2Y

In contrast to the traditional role of national parks of showcasing rare biological and geological phenomena, the 1974 MAB "criteria and guidelines" called for the biosphere reserve network to represent the world's full biogeographic diversity—a diversity reflected in the inconsistent terminological variations of "biogeographic realms," "biogeographic provinces," "biomes," "biotic divisions," and "characteristic ecosystems" (Franklin 1977, 263; Hough 1988, 1; Tangley 1988, 148). Accordingly, the MAB Program sponsored the development of a global biogeographical classification system authored by Udvardy (1975). After the 1970s, however, Udvardy's system became outdated as a conservation tool, and the goal of complete representation appears to have diminished somewhat as a MAB Program priority (Magin and Chape 2004; UNESCO 2001, 20).

A more influential goal under the 1974 criteria and guidelines was a particular type of zoning system for individual biosphere reserves that distinguished between "core" and "buffer" areas (Figure 2.1). Although conservationists had used the language and concepts of buffer zones decades before (Shafer 1999; Shelford 1933), this multitiered land management structure stood in marked contrast to most previous conceptualizations of protected areas, where nearly all designated land was set off-limits to human use. This zoning system became "a key element of the biosphere reserve concept" and the "dominant model for biosphere reserves during the 1970s" (Hough 1988, 1).

In the 1980s, another layer, alternatively called a transition or cooperation zone, was added to the zoning system (Hough 1988, 1). Other land classifications were subsequently integrated into this model, including manipulative research areas, rehabilitation areas, human settlement areas,

> **Fig. 2.1**
> *Examples of Zoning Systems for Biosphere Reserves*
>
> ```
>         The following examples illustrate the "core" and "buffer" zone concepts.
> The diagrams are simplified but they suggest a few of the complex situations in
> different biomes.
>
> 1.   Basic "core" and "buffer" zones and possible uses of these zones.
>
>      core     - no development permitted ; uses are strictly controlled.
>
>      buffer 1 - used for research and educational purposes ; use and movement
>                 by the public are limited to authorized sections and tracks.
>
>      buffer 2 - may be used for various purposes, including public recreation,
>                 but uses are controlled according to the carrying capacity of
>                 the area.
> ```
>
> [diagram of concentric zones: core inside buffer 1 inside buffer 2]

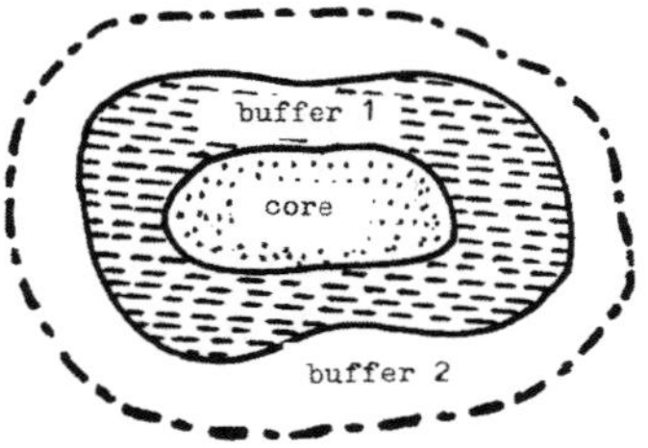

and traditional use areas (Figure 2.2 is but one modern representation of the system). Because site-specific conditions and local constraints would often alter the idealized system of concentric rings, MAB early on adopted a strategy of "cluster reserves" or "multiple reserves" that apportioned different zones in discrete areas (Batisse 1997; Franklin 1977; Hough 1988).

As a "radically new" idea in the early 1970s, the biosphere reserve model was considered a "major achievement" in the global conservation movement and "the best suited instrument" for achieving both conservation and sustainable development (de Klemm and Shine 1993, 271; Dyer and Holland 1991, 319; Ezcurra et al. 1999, 174). Yet few biosphere reserves ever fit the ideal model or even its variants. Batisse (1997, 31) generously explained that the implementation of the biosphere reserve concept "developed as a result of progressive adaptation to field realities and experience rather than any well-designed, rigidly formulated, and preconceived plan." Dyer and Holland (1991) argued more forcefully that although the biosphere reserve concept was "well ahead of its time" when originally conceived, it "cannot address the major ecological issues of the times." And indeed, as the system expanded through the late 1970s and 1980s, it

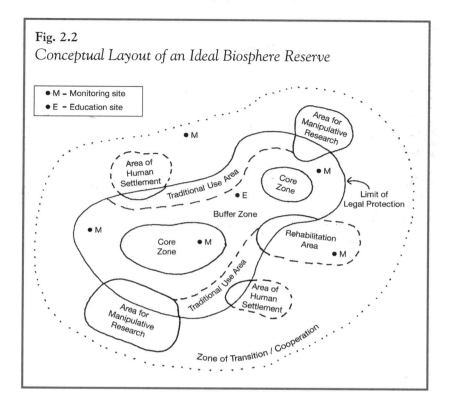

Fig. 2.2
Conceptual Layout of an Ideal Biosphere Reserve

came to be seen more as a label than a land-use system, and the zoning system was characterized as a "myth" with "little difference between the management of biosphere reserves and the underlying protected areas" (Hough 1988, 2). Such critiques were substantiated by an IUCN study revealing that 84 percent of existing biosphere reserves overlaid preexisting protected areas—a finding that led others to criticize MAB's disproportionate focus on its conservation mission over its development and logistics missions (Batisse 1997; Gregg 1991; Tangley 1988). "Once again," Adams (1990, 36) gloomily reflected, "a 'global' initiative generated in the North is reduced to rather small ripples on the periphery."

Despite hopes that the biosphere reserve network would serve as a "pillar upon which to build broad, regional-land-management plans," the IUCN study also demonstrated that—at least in terms of geographic extent—biosphere reserves were generally far more aligned with national parks than with bioregions (Tangley 1988, 150). Nevertheless, over the course of the 1980s, the basic zoning model promoted for biosphere reserves was increasingly integrated into various manifestations of bioregional thinking. As Batisse (1997) put it, the biosphere reserve concept was "progressively finding its place in the larger framework of bioregional ecosystem management and land-use planning."

More specifically, the biosphere reserve system held up a tangible model for conservation biologists and others who were delving into the relatively unexplored subjects of landscape ecology, habitat fragmentation, reserve networks, wildlife corridors, and habitat connectivity. Although ridden at times with rancorous debate, these intellectual endeavors collectively put forth one simple underlying tenet for land conservation: *Think big and be inclusive*. Even if the MAB Program had not sufficiently focused on thinking big (though recall di Castri's emphasis on Earth as *the* biosphere reserve), it had generated widespread understanding of the significance of incorporating both core areas and buffer zones in land-use design and management. It was from within that intellectual zeitgeist, as will be described at much greater length in Chapter 4, that Y2Y would arise.

Biosphere Reserves, Local Participation, and ISDA

"On paper, biospheres look wonderful," said former ISDA coordinator Ceal Smith, but "most reserves have problems because they have not succeeded in integrating, in any meaningful way, the needs of communities." Smith was spelling out a point made by her former professor Raymond Dasmann (1988, 489), who had asserted that the long-term viability of MAB's biosphere reserve model depended on "local support or at least acceptance." By the mere fact of incorporating buffer zones as an integral component of a biosphere reserve, the MAB Program had implicitly accepted that basic principle from the start. Yet as an operational principle, garnering "local participation" in the establishment and operation of biosphere reserves had not been part of the "conceptual base" of MAB's original plan (Halffter and Ezcurra 1987, 192).

By the 1980s, the lack of attention to local participation had become recognized as a significant problem in establishing new biosphere reserves (Halffter and Ezcurra 1987). The problem was addressed in the 1984 *Action Plan for Biosphere Reserves*, a tenth-anniversary updating of the 1974 criteria and guidelines. One of the fundamental objectives outlined in the *Action Plan* was to obtain consent and active support from all regional stakeholders in the processes of planning and management (Batisse 1985, 1997).

This emphasis on local participation played a fundamental role in the development of ISDA. Indeed, ISDA participants largely interpreted the biosphere concept as "a framework for regional cooperation involving government decision-makers, scientists, resource managers, private organizations, and local people" (Murrieta-Saldivar 1996, 2). As such, ISDA was able to draw in a much broader network of interests than it would have if MAB had been interpreted to be solely concerned with land classification and management. But as Chapter 3 will make clear, ISDA's interpretation of the biosphere reserve concept also entailed the challenge of surmounting tremendous cross-cultural differences within the Sonoran Desert. In addition, it ran headlong into a *very* different interpretation of biosphere reserves, an interpretation that regarded them as part of a vast international conspiracy to defraud the United States of sovereignty over its own lands. Ultimately, such conspiratorial thinking foiled ISDA's efforts to create a truly *bioregional* biosphere reserve.

ISDA and Y2Y: New Forms of Transborder Conservation

To borrow from the somewhat tepid language of the biological sciences, transborder conservation occurs across multiple temporal and spatial scales. This diversity is reflected *within* both the Sonoran Desert and the Y2Y region. Crossing their respective borders, for example, the Sonoyta and Flathead Rivers have both received attention from conservationists (Environment Committee of the Arizona-Mexico Commission 1988; Reardon 1992). At a larger scale, the Y2Y region contains the world's first international peace park, and work continues on establishing such an entity in the Sonoran Desert (Sifford and Chester, forthcoming). Although no "transboundary biosphere reserves" have been officially designated in either the Y2Y region or the Sonoran Desert (only six such entities exist, five in Europe and one in Africa), both regions contain biosphere reserves adjacent to both sides of the border: the Waterton Reserve next to the Glacier Biosphere Reserve on the northern border, and the Organ Pipe Biosphere Reserve next to the Pinacate and Alto Golfo Biosphere Reserves on the southern border (UNESCO 2003; UNESCO-MAB 2005).

Measured by their geographic scope alone, both ISDA and Y2Y constitute a wholly different and still relatively new type of international entity.

Both cover areas largely equivalent to or much larger than the land classification category of bioregion or ecoregion, which have been loosely approximated at 100,000 square kilometers (or about 39,000 square miles) (Bailey 1996, 24). Only a handful of the larger TBPA efforts—such as the proposed Great Limpopo Transfrontier Park in southern Africa at 35,000 square kilometers (13,650 square miles), envisioned to eventually encompass a "transfrontier conservation area" of 100,000 square kilometers (39,000 square miles)—equal the areas covered by ISDA at 96,000 and Y2Y at 1,200,000 square kilometers (37,440 and 468,000 square miles, respectively; Department of Environment and Tourism n.d.). Although the two developed in dramatically different fashions, they share several other characteristics:

- Both ISDA and Y2Y are composed mostly of civil society actors, including indigenous peoples, unaffiliated citizens, and activists; in contrast, government officials have been involved to a much lesser degree.
- Participants in both initiatives seek to protect the biodiversity of the region and to promote sustainable economic development.
- Much of the land in each region is owned and managed by federal, state, provincial, and indigenous governments. Consequently, in addition to their respective international borders, ISDA and Y2Y cross a complex web of federal, state, provincial, and indigenous borders.
- Both began as loosely organized entities that over time became more formalized, which entailed the establishment of boards of directors and central offices with dedicated, full-time, paid staff.

Although those similarities may keep ISDA and Y2Y in the same taxonomic family, a comparison of Chapters 3 and 4 will demonstrate that they are not the same species.

In sum, both ISDA and Y2Y are far more than attempts to create "nature parks" writ large. But while the question of what they *are* is of more than academic interest, conservationists will be most interested in how effective they have been. The following two chapters provide extensive answers to both questions. Chapter 5 will then approach the latter question of effectiveness from a comparative perspective.

References

Adams, W. M. 1990. *Green development*. New York: Routledge.

African Wildlife Federation. 2005. The Democratic Republic of the Congo, Rwanda and Uganda unite for conservation in the Central Albertine Rift. October 24. http://www.awf.org/news/31506.

Bailey, Robert G. 1996. *Ecosystem geography*. New York: Springer.

Bailey, Robert G. 2002. *Ecoregion-based design for sustainability*. New York: Springer.

Bakarr, Mohamed Imam. 2003. Conservation on the frontier. *Tropical Forest Update*: 3–5. http://www.itto.or.jp/live/contents/download/tfu/TFU.2003.02.English.pdf.

Batisse, Michel. 1985. Action plan for biosphere reserves. *Environmental Conservation* 12, no. 1: 17–27.

Batisse, Michel. 1997. Biosphere reserves: A challenge for biodiversity conservation and regional development. *Environment* (June): 6–15, 31–33.

Benvenisti, Eyal. 2002. *Sharing transboundary resources: International law and optimal resource use*. New York: Cambridge University Press.

Berkman, Paul Arthur. 2003. International Polar Year 2007–08. *Science* 301, no. 5640: 1669.

Besançon, Charles, and Anne Hammill. Forthcoming. Measuring peace park performance: Transboundary mountain gorilla conservation in central Africa. In *Peace parks: Conservation and conflict resolution*, ed. Saleem Ali. Cambridge, MA: MIT Press.

Besançon, Charles, and Conrad Savy. 2005. Global list of internationally adjoining protected areas and other transboundary conservation initiatives. In *Transboundary conservation: A new vision for protected areas*, ed. Russell A. Mittermeier, Patricio Robles Gil, Christina G. Mittermeier, Cyril Kormos, Trevor Sandwith, and Charles Besançon. Washington, DC: Conservation International, Agrupacion Sierra Madre, and CEMEX.

Birnie, Patricia W., and Alan E. Boyle. 1992. *International law and the environment*. New York: Clarendon Press; Oxford University Press.

Brosius, J. Peter, and Diane Russell. 2003. Conservation from above: An anthropological perspective on transboundary protected areas and ecoregional planning. In *Transboundary protected areas: The viability of regional conservation strategies*, ed. Urami Manage Goodale, 39–65. New York: Food Products Press (published simultaneously in the *Journal of Sustainable Forestry*, 17: 1–2).

Brunnée, Jutta. 2004. The United States and international environmental law: Living with an elephant. *European Journal of International Law* 15, no. 4: 617–649.

Brunner, Robert. 1999. *Transboundary protected areas in Europe*. Ljubljana, Slovenia: EUROPARC Federation (FNNPE) and World Commission on Protected Areas of the World Conservation Union. November.

Canadian Embassy. 2001. *United States–Canada: The world's largest trading relationship*. Washington, DC: Canadian Embassy. http://www.canadianembassy.org/trade/wltr2001-en.pdf.

Commission on Environmental Law. 2004. *Draft international covenant on environment and development* (3rd ed., updated text). Prepared in cooperation with the International Council of Environmental Law. Cambridge, UK: Author and World Conservation Union (IUCN).

Conca, Ken, and Geoffrey D. Dabelko. 2003. *Environmental peacemaking.* Washington, DC: Woodrow Wilson Center Press.

Convention on Biological Diversity. 2004. Annex: Decisions adopted by the Conference of the Parties to the Convention on Biological Diversity at its seventh meeting. UNEP/CBD/COP/7/21 (conference document). http://www.biodiv.org/doc/decisions/COP-07-dec-en.pdf.

Council of Europe. European Outline Convention on Trans-frontier Co-operation between Territorial Communities or Authorities (Madrid, 21.V.1980). http://conventions.coe.int/Treaty/EN/Treaties/Html/106-1.htm [with link to Model Agreement].

Dasmann, R. F. 1966. Man in North America. In *Future environments of North America: Being the record of a conference convened by the Conservation Foundation in April, 1965, at Airlie House, Warrenton, Virginia,* ed. F. Fraser Darling and John P. Milton, 326–334. Garden City, NY: Natural History Press.

Dasmann, Raymond. 1973–74. Conservation of natural resources. In *The dictionary of the history of ideas: Studies of selected pivotal ideas,* ed. Philip P. Wiener, vol. 1, 470–477. New York: Charles Scribner's Sons.

Dasmann, Raymond F. 1988. Biosphere reserves, buffers, and boundaries. *BioScience* 38, no. 7: 487–489.

Dasmann, Raymond Fredric. 2002. *Called by the wild: The autobiography of a conservationist.* Berkeley: University of California Press.

Department of Environmental Affairs and Tourism. n.d. *Great Limpopo Transfrontier Park.* Department of Environmental Affairs and Tourism, South African government. http://www.environment.gov.za/projprog/tfcas/limpopo/index_limpopo.htm.

de Klemm, Cyrille, and Clare Shine. 1993. *Biological diversity conservation and the law.* Gland, Switzerland: IUCN.

Dorsey, Kurkpatrick. 1998. *The dawn of conservation diplomacy: U.S.-Canadian wildlife protection treaties in the progressive era.* Seattle: University of Washington Press.

Douglas, Ross. 1997. Parks, peace and prosperity. *Africa—Environment and Wildlife* (September-October), 30–39.

Draper, Malcolm, Marja Spierenburg, and Harry Wels. 2004. African dreams of cohesion: Elite pacting and community development in transfrontier conservation areas in southern Africa. *Culture and Organization* 10, no. 4: 341–353.

Dutton, Sheila, and Fiona Archer. 2004. Transfrontier parks in South Africa. *Cultural Survival Quarterly* 28; no. 1 (30 April), 57–60.

Dyer, M. I., and M. M. Holland. 1991. The biosphere-reserve concept: Needs for a network design. *BioScience* 41: 319–325.

Environment Committee of the Arizona-Mexico Commission, ed. 1988. *Simposio de Investigación sobre la Zona Ecológica de El Pinacate.* Hermosillo, Sonora, Mexico.

EUROPARC Federation. 2003. Transboundary parks. . . . following nature's design. *EUROPARC Newsletter* no. 5 (August), 2–3.

Ezcurra, Exequiel, Marisa Mazari-Hiriart, Irene Pisanty, and Adrián Guillermo Aguilar. 1999. *The basin of Mexico: Critical environmental issues and sustainability*. New York: United Nations University Press.

Fall, Juliet J. 2005. *Drawing the line: Nature, hybridity, and politics in trans-boundary spaces*. Burlington, Vermont: Ashgate.

Fall, Juliet J., and Giorgio Andrian. 2005. Experiment and adjust: Tracing changes in the definition of boundaries in biosphere reserves. *Rivista Gazzetta Ambiente*, no. 4: 23–32.

Fitter, Richard, Sidney Richmond, and Peter Markham Scott. 1978. *The penitent butchers: The Fauna Preservation Society, 1903–1978*. London: F.P.S.

Fossey, Dian. 1983. *Gorillas in the mist*. Boston: Houghton Mifflin.

Franklin, Jerry F. 1977. The biosphere reserve program in the U.S. *Science* 195: 262–267.

Franklin, Jerry. 1999. Foreword. In *Bioregional assessments: Science at the crossroads of management and policy*, ed. K. Norman Johnson, Frederick Swanson, Margaret Herring, and Sarah Green, xi–xiii. Washington, DC: Island Press.

Gartlan, Steve. 1999. The Central African experience in transfrontier protected areas. In *Partnerships for protection: New strategies for planning and management for protected areas*, ed. Sue Stolton and Nigel Dudley, 243–248. London: Earthscan.

Global Environment Facility. 1996. *Mozambique: Transfrontier Conservation Areas Pilot and Institutional Strengthening Project*. GEF Project Document 15534. Washington, DC: Author and World Bank. http://www-wds.worldbank.org/servlet/WDS_IBank_Servlet?pcont=details&eid=000009265_3970311113856.

Global Transboundary Protected Areas Network. n.d. *Transboundary protected areas*. Gland, Switzerland: World Conservation Union (IUCN). http://www.tbpa.net/index.html.

Goetel, Walery. 1964. Parks between countries. In *First World Conference on National Parks: Proceedings of a conference organized by the International Union for Conservation of Nature and Natural Resources (IUCN)*, ed. Alexander B. Adams, 287–294. Washington, DC: National Park Service, U.S. Department of the Interior.

Golley, Frank B. 1993. *A history of the ecosystem concept in ecology: More than the sum of the parts*. New Haven, CT: Yale University Press.

Gregg, William P. Jr. 1991. MAB biosphere reserves and conservation of traditional land use systems. In *Biodiversity: Culture, conservation, and ecodevelopment*, ed. Margery L. Oldfield and Janis B. Alcorn, 274–294. Boulder, CO: Westview Press.

Gregg, William P. 1999. Environmental policy, sustainable societies and biosphere reserves. In *Ecosystem management for sustainability: Principles and practices illustrated by a regional biosphere reserve cooperative*, ed. John D. Peine, 23–40. Boca Raton, FL: Lewis Publishers.

Griffin, John, David Cumming, Simon Metcalfe, Mike t'Sas-Rolfes, Jaidev Singh, Ebenizário Chonguiça, Mary Rowen, and Judy Oglethorpe. 1999. *Study on the development of transboundary natural resource management areas in southern Africa: Main report*. Publication number 65. Washington, DC: Biodiversity Support Program. http://www.worldwildlife.org/bsp/publications/africa/trans_main/griffin-01.html.

Halffter, Gonzalo, and Exequiel Ezcurra. 1987. Evolution of the biosphere reserve concept. In *Proceedings of the Symposium on Biosphere Reserves, Fourth World Wilderness Congress, September 14–17, 1987, Estes Park, Colorado*, ed. William P. Gregg Jr., Stanley L. Krugman, and James D. Wood Jr., 188–206. Atlanta, GA: U.S. National Park Service, Department of the Interior.

Hamilton, Lawrence S. 2001. International transboundary cooperation: Some best practice guidelines. In *Crossing boundaries in park management: Proceedings of the 11th Conference on Research and Resource Management in Parks and on Public Lands*, ed. David Harmon, 34. Hancock, MI: George Wright Society.

Hamilton, L. S., J. C. Mackay, G. L. Worboys, R. A. Jones, and G. B. Manson. 1996. *Transborder protected area cooperation*. Canberra: Australian Alps Liaison Committee and World Conservation Union.

Harris, Elizabeth, Chase Huntley, William Mangle, and Naureen Rana. 2001. *Transboundary collaboration in ecosystem management: Integrating lessons from experience*. University of Michigan, School of Natural Resources & Environment. http://www.snre.umich.edu/emi/pubs/transboundary.htm.

Harrison, J. M. 1982. MAB: A ten-year perspective. Ecological Monitoring and Assessment Network Coordinating Office, Environment Canada. Last accessed September 2001 at http://www.cciw.ca/mab/reports/newsletters/communique_14/art2e.html; currently not available.

Hayes, Peter. 2001. *Environmental security in a world of perpetual war*. Presented at Environmental Grantmakers Association, Brainerd, Minnesota, October 8. Nautilus Institute. http://www.nautilus.org/archives/papers/security/EnvironmentalSecurity-War.pdf.

Heijnsbergen, P. van. 1997. *International legal protection of wild fauna and flora*. Washington, DC: IOS Press.

Henderson, John Brooks. 1901. *American diplomatic questions*. New York: Macmillan.

Herman, Otto. 1907. *The International Convention for the Protection of Birds concluded in 1902; and Hungary*. Budapest, Hungary: V. Hornyánszky.

Holdgate, Martin. 1999. *The green web*. London: Earthscan.

Homer-Dixon, Thomas F. 1999. *Environment, scarcity, and violence*. Princeton, NJ: Princeton University Press.

Hough, John. 1988. Biosphere reserves: Myth and reality. *Endangered Species Update*, 6, nos. 1 and 2: 1–4.

Hughes, Ross, and Fiona Flintan. 2001. *Integrating conservation and development experience: A review and bibliography of the ICDP literature*. Cork, Ireland: University College Cork. http://www.ucc.ie/famine/GCD/ICDP_sec.pdf.

International Tropical Timber Organization. 2004. *ITTO-funded transboundary and other conservation area initiatives*. http://www.itto.or.jp/live/PageDisplayHandler?pageId=59.

Jarrell, Randall. 2000. *Raymond F. Dasmann: A life in conservation biology*. Santa Cruz: University of California.

Kahn, Tamar. 2003. *Breaking down borders in Africa*. SciDev.Net. http://www.scidev.net/Features/index.cfm?fuseaction=readFeatures&itemid=200&language=1.

Kukura, Rudolf. n.d. *Pieniny*. http://www.pieniny.sk/en.html.

Lanjouw, A., A. Kayitare, H. Rainer, E. Rutagarama, M. Sivha, S. Asuma, and J. Kalpers. 2001. *Beyond boundaries: Transboundary natural resource management for mountain gorillas in the Virunga-Bwindi Region*. Washington, DC: Biodiversity Support Program. http://www.worldwildlife.org/bsp/publications/africa/126/titlepage.HTML.

Lee, Hye-Jong Linda. 2004. The pragmatic Migratory Bird Treaty Act: Protecting "property." *Boston College Environmental Affairs Law Review*, 31, no. 3: 649–681.

Louka, Elli. 2002. *Biodiversity & human rights: The international rules for the protection of biodiversity*. Ardsley, NY: Transnational Publishers.

Lyster, Simon. 1985. *International wildlife law*. New York: Cambridge University Press.

MacDonald, Graham A. 2000. *Where the mountains meet the prairies: A history of Waterton County*. Calgary, Canada: University of Calgary Press.

Mackinnon, Kathy. 2002. World Bank assistance to transboundary reserves. *Oryx* 36, no. 2: 115.

Magin, Chris, and Stuart Chape. 2004. *Review of the world heritage network: Biogeography, habitats, and biodiversity*. United Nations Environment Program, World Conservation Monitoring Centre, and World Conservation Union. July. http://www.unep-wcmc.org/protected_areas/world_heritage/Introduction_WHN.pdf.

Merchant, Carolyn. 2002. *The Columbia guide to American environmental history*. New York: Columbia University Press.

Milich, Lenard, and Robert G. Varady. 1998. Managing transboundary resources: Lessons from river-basin accords. *Environment* (October): 10–15, 35–41.

Mulongoy, Kalemane Jo, and Stuart P. Chape. 2004. *Protected areas and biodiversity: An overview of key issues*. Montreal, Canada, and Cambridge, UK: Secretariat of the Convention on Biological Diversity, UN Environment Program, World Conservation Monitoring Center. http://www.biodiv.org/doc/publications/pa-brochure-en.pdf.

Murrieta-Saldivar, Joaquin. 1996. *Community workshops in the Sonoran Desert border region*. Ajo and Tucson, AZ: International Sonoran Desert Alliance, Sonoran Institute.

Poore, Duncan. 2003. *Changing landscapes: The development of the International Tropical Timber Organization and its influence on tropical forest management*. Sterling, VA: Earthscan.

Reardon, Carol. 1992. The International Joint Commission: A possible role model for international resource management. In *International environmental treaty making*, ed. Lawrence Susskind, Eric Jay Dolin, and J. William Breslin, 125–142. Cambridge, MA: Harvard Law School Program on Negotiation.

Reyers, Belinda. 2003. *Evaluating transboundary protected areas: Achieving biodiversity targets*. Paper prepared for the Workshop on Transboundary Protected Areas in the Governance Stream of the 5th World Parks Congress, Durban, South Africa, September 12–13. http://www.tbpa.net/docs/WPCGovernance/BelindaReyers.pdf.

Rikoon, J. Sanford, and Theresa L. Goedeke. 2000. *Anti-environmentalism and citizen opposition to the Ozark Man and the Biosphere Reserve*. Lewiston, NY: Edwin Mellen Press.

Rogers, Peter J. 2002. *Global governance/governmentality, wildlife conservation, and protected area management: A comparative study of eastern and southern Africa.* Paper presented at the African Studies Association 45th Annual Meeting, Washington DC, December 5. http://hdgc.epp.cmu.edu/misc/Rogers2002.doc.

Rotary International, Glacier National Park, and Waterton Lakes National Park. n.d. *It began as a bold idea.* http://www.nps.gov/glac/pdf/rotary_web.pdf.

Rüster, Bernd, and Bruno Simma, eds. 1975. *Treaties and related documents.* Vol. 4 of International protection of the environment. Dobbs Ferry, NY: Oceana.

Sale, Kirkpatrick. 1985. *Dwellers in the land: The bioregional vision.* San Francisco: Sierra Club Books.

Sandwith, Trevor, and Charles Besançon. 2005. *Report on the activities of the Task Force in 2004.* Claremont, South Africa: Task Force on Transboundary Protected Areas, World Commission on Protected Areas, IUCN.

Sandwith, Trevor, Clare Shine, Lawrence Hamilton, and David Sheppard. 2001. *Transboundary protected areas for peace and co-operation.* Gland, Switzerland: World Conservation Union.

Shafer, Craig L. 1999. US National Park buffer zones: Historical, scientific, social, and legal aspects. *Environmental Management* 23, no. 1: 49–73.

Shelford, Victor E. 1933. The preservation of natural biotic communities. *Ecology Letters* 14, no. 2: 240–245.

Sifford, Belinda, and Charles C. Chester. Forthcoming. 2006. Bridging conservation across La Frontera: An unfinished agenda for peace parks along the U.S.-Mexico divide. In *Peace parks: Conservation and conflict resolution,* ed. Saleem Ali. Cambridge, MA: MIT Press.

Sochaczewski, Paul Spencer. 1999. Across a divide. *International Wildlife* 29, no. 4 (July–August): 34–41.

Spencer, Robert, John J. Kirton, and Kim Richard Nossal, eds. 1981. *The International Joint Commission seventy years on.* Toronto, Canada: Centre for International Studies, University of Toronto.

Stewart, Bob. n.d. *Peace, parks and Rotary.* Okotoks, Alberta: Rotary Club of Okotoks. http://www.peace.ca/rotarypeaceparks.htm.

Stone, Richard, and Gretchen Vogel. 2004. Polar exploration: A year to remember at the ends of the earth. *Science* 303, no. 5663: 1458–1461.

Tangley, Laura. 1988. A new era for biosphere reserves. *BioScience* 38, no. 3: 148–155.

Thorsell, Jim, and J. Harrison. 1990. Parks that promote peace: A global inventory of transfrontier nature reserves. In *Parks on the borderline: Experience in transfrontier conservation,* ed. Jim Thorsell, 3–21. Gland, Switzerland: World Conservation Union.

Times of London. 1900. Preservation of wild animals in Africa. May 30: 12. Times Digital Archive, http://www.galegroup.com/Times/.

Tolentino, Amado S. Jr. 2001. A call for a legal regime on transborder parks. *International Review for Environmental Strategies* 2, no. 1: 43–60.

Transboundary Protected Areas Task Force. 2004. *Workshop report: Transboundary Task Force meeting on La Maddalena Island (May 17–22).* World Conservation

Union and World Commission on Protected Areas. http://www.iucn.org/themes/wcpa/theme/parks/LaMadallenaMtgSummary.pdf.

Udvardy, Miklos D. F. 1975. *A classification of the biogeographical provinces of the world*. IUCN Occasional Paper No. 18. Morges, Switzerland: International Union for Conservation of Nature and Natural Resources.

UNESCO. 1974a. *International Co-ordinating Council of the Programme on Man and the Biosphere (MAB), Third session, Final report*, Washington DC, 17–29 September. Paris. MAB report series No. 27.

UNESCO. 1974b. *Programme on Man and the Biosphere Task Force on: Criteria and guidelines for the choice and establishment of biosphere reserves*. May 20–24. MAB report series No. 22. Paris: UNESCO and United Nations Environment Programme.

UNESCO. 2001. *Seville + 5 International Meeting of Experts (Proceedings, Pamplona, Spain, 23–27 October 2000)*. SC-2001/WS/28. Paris: Man and the Biosphere Programme. http://unesdoc.unesco.org/images/0012/001236/123605m.pdf.

UNESCO. 2003. *Five transboundary biosphere reserves in Europe*. Biosphere Reserves Technical Notes. Paris: Man and the Biosphere Programme.

UNESCO. 2005a. *International Coordinating Council of the Man and the Biosphere (MAB)*. General Conference, 33rd session: 3–21 October. http://portal.unesco.org/en/ev.php-URL_ID=10048&URL_DO=DO_TOPIC&URL_SECTION=201.html.

UNESCO. 2005b. Michel Batisse (1923–2004). http://www.unesco.org/mab/news/MB/obituary.pdf.

UNESCO-MAB Secretariat. 2005. *Biosphere Reserves World Network*. Paris: Division of Ecological and Earth Sciences. June. http://www.unesco.org/mab/brlist.PDF.

UNESCO-MAB. 2005. *Benin/Burkina Faso/Niger: 'W' Region*. http://www2.unesco.org/mab/br/brdir/directory/biores.asp?code=BEN+-+BKF+-+NER+01&mode=all.

United Nations Environment Programme, World Conservation Monitoring Center, and World Commission on Protected Areas. 2004. *Transboundary protected areas*. Protected Aeas and World Heritage Programme. http://www.unep-wcmc.org/index.html?http://www.unep-wcmc.org/protected_areas/transboundary/~main.

U.S. Fish and Wildlife Service. 1991. *National Wildlife Refuge System (Organization and History Series: 029 FW 3)*. Division of Policy and Directives Management, U.S. Fish and Wildlife Service. http://www.fws.gov/policy/029fw3.html.

U.S. Fish and Wildlife Service. n.d. *Alaska Maritime National Wildlife Refuge: Area history*. http://alaska.fws.gov/nwr/akmar/historyculture/1800s.htm.

U.S. Fish and Wildlife Service and Direction General for Conservation and Ecological Use of Natural Resources of Mexico. n.d. *Sixty years of cooperation between the United States and Mexico in biodiversity conservation (1936–1996)*. Prado Norte, Federal District, Mexico: Authors.

Weber, William, and Amy Vedder. 2001. *In the kingdom of gorillas: Fragile species in a dangerous land*. New York: Simon & Schuster.

Westing, Arthur H. 1993a. Biodiversity and the challenge of national borders. *Environmental Conservation* 20, no. 1: 5–6.

Westing, Arthur H. 1993b. Building confidence with transfrontier reserves: The global potential. In *Transfrontier reserves for peace and nature: A contribution to human security*, ed. Arthur H. Westing, 1–15. Nairobi, Kenya: United Nations Environment Programme.

Westing, Arthur H. 1998. Establishment and management of transfrontier reserves for conflict prevention and confidence building. *Environmental Conservation* 25, no. 2: 91–94.

Wilcher, Marshall E. 1980. *Environmental cooperation in the North Atlantic area.* Washington, DC: University Press of America.

Wilkie, D. S., E. Hakizumwami, N. Gami, and B. Difara. 2001. Beyond boundaries: Regional overview of transboundary natural resource management in Central Africa. In *Beyond boundaries: Transboundary natural resource management in Central Africa.* Washington, DC: Biodiversity Support Program. http://www.worldwildlife.org/bsp/publications/africa/125/125/titlepage.HTML.

Williams, Ted. 1997. The baiting game. *Audubon* (May-June): 28.

Wolf, Aaron T., Jeffrey A. Natharius, Jeffrey J. Danielson, Brian S. Ward, and Jan K. Pender. 1999. International river basins of the world. *International Journal of Water Resources Development* 15, no. 4. http://www.transboundarywaters.orst.edu/publications/register/register_paper.html.

World Commission on Protected Areas. 2003. *WCPA Task Force: Transboundary protected areas.* World Conservation Union. http://www.iucn.org/themes/wcpa/theme/parks/parks.html.

Worthington, E. Barton. 1975. *The evolution of IBP.* New York: Cambridge University Press.

Zbicz, Dorothy C. 1999. Transboundary cooperation between internationally adjoining protected areas. In *On the frontiers of conservation: Proceedings of the 10th Conference on Research and Resource Management in Parks and on Public Lands, Asheville, NC, 22–26 March 1999*, ed. David S. Harmon. Hancock, MI: George Wright Society, 199–204.

Zbicz, Dorothy C. 2003. Imposing transboundary conservation: Cooperation between internationally adjoining protected areas. In *Transboundary protected areas: The viability of regional conservation strategies*, ed. Urami Manage Goodale, Marc J. Stern, Cheryl Margoluis, Ashley G. Lanfer, and Matthew Fladeland, 21–37. New York: Food Products Press (published simultaneously in the Journal of Sustainable Forestry, 17: 1–2).

3

Border Biosphere Reserves and the International Sonoran Desert Alliance

The question went something like this: "Where exactly are we talking about?" It was a question of foundational significance to the eclectic 1993 gathering of Native Americans, community activists, scientists, conservationists, business people, and a handful of government officials. Having assembled several times over the previous year with the purpose of addressing the region's social and environmental ills, this group collectively felt the need to come to some basic agreement over *where* they all were.

Notably, even before posing the question, they had started out with a label for their chosen geography: the western Sonoran Desert. But what that encompassed was up for grabs. After lengthy discussion of the region's boundaries (there was always somebody asking, "Why stop there?"), a soft-spoken official from the U.S. Bureau of Land Management named Beau McClure offered to go back to his office and draw up a map.

McClure came back with a circle drawn around a sparsely inhabited area largely composed of federal and tribal lands lying south and west of Arizona's two largest cities, Phoenix and Tucson (CCR 1993) (Figure 3.1). The map marked an important development for the group, almost as significant as the nearly contemporaneous adoption of a permanent title: the International Sonoran Desert Alliance. Although the label of "western" Sonoran Desert has since fallen out of use, ISDA has consistently focused on that 96,000-square-kilometer (37,440-square-mile) spread of land to the present day.

It may be just as well that the term has fallen out of use, for when viewed from the perspective of the entire Sonoran Desert, which generally surrounds the Gulf of California in an inverted U, the area lies at the "top" of the region—thus tempting one to call it the northern Sonoran Desert. But the designation makes sense from the perspective of the culturally cohesive Arizona-Sonora borderlands, a region that the Spaniards called alternatively *la Pimería Alta* (land of the Northern Pima) and *la Papaguería* (land of the Papagos) (Broyles et al. 1997). Because the region's major population centers lie well within its eastern half, the designation of "western" is not so mysterious.

To understand how ISDA formally came into being in 1993, we will pick up the thread begun in Chapter 2 on the MAB Program. Specifically, we will look at national implementation of the MAB Program in both Mexico and the United States in order to understand the "long and interesting record of efforts to implement biosphere reserve concepts" in the Sonoran Desert (Gregg 1992). First, however, it is necessary to get a sense of the lay of the land—to understand "where exactly we are talking about."

Contemporary Land Management Patterns in the Western Sonoran Desert

Driving along many of the Sonoran Desert's major roads and highways—away from which few visitors and only few inhabitants wander—the surrounding land appears mountainous (the highest point being Ajo Peak at 1,454 meters, or 4,770 feet) (U.S. National Park Service 1965, 10). Yet maps and aerial views show that the vast majority of the area is flat desert, with intermittent parallel north-south–trending mountain ranges spiking up as if clawed from underneath by a goliath subterranean *león de la sierra*.

This is the land of the O'odham, an Uto-Aztecan speaking people whose origins remain contested among archaeologists and anthropologists (Sheridan and Parezo 1996, 116–117). But whether they had been on the land for thousands or only a couple hundred years, the land was their home when Cabeza de Vaca skirted the area in the 1530s. In the subsequent 150 years only a small number of European explorers would curse the pounding heat of the western Sonoran Desert. European culture did not establish a foothold in the region until the 1687 arrival of Jesuit priest Father Eusebio Kino, whose peripatetic wanderings over la Pimería Alta (his term) led to the establishment of over twenty Catholic missions (Polzer 1998, 129–130).

Through a labyrinthine historical evolution from territorial control to independence during the eighteenth and first half of the nineteenth centuries, the western Sonoran Desert was nominally controlled by a series of administrative structures emanating from New Spain and then Mexico. Even after the U.S.-Mexico War (1846–48) and the 1848 Treaty

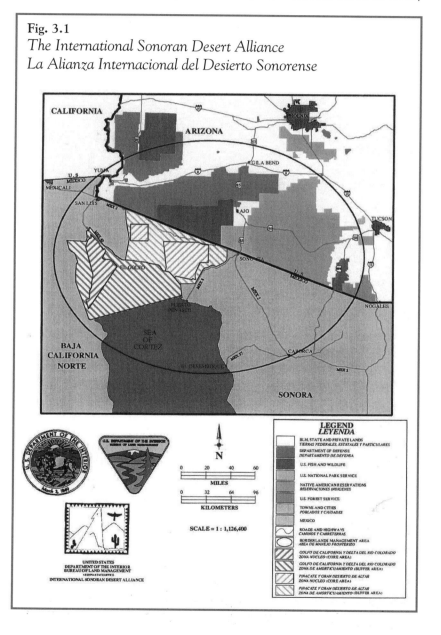

Fig. 3.1
The International Sonoran Desert Alliance
La Alianza Internacional del Desierto Sonorense

of Guadalupe Hidalgo, the area remained mostly within Mexico (Beck and Haase 1989, Section 41). Only in 1853, when the recently installed dictatorial *Serenissima Altera* Santa Anna found his government bankrupt, did Mexico sell a large tract of land to the United States through the Gadsden Treaty (ratified in 1854). Covering what would become approximately the lower third of Arizona and the southwestern corner of New Mexico, this new territory of the United States was bounded to the south by a new international border running 558 kilometers (346 miles) between Nogales on the east and Mexicali on the west. This unyieldingly straight line sliced in half not only the western Sonoran Desert, but also the O'odham culture living within (Guia Roji 1998, 55).

The western Sonoran Desert is today administratively divided into several extremely large land management units. On the U.S. side of the border, the Tohono O'odham Nation reservation, established in 1916, constitutes the second second-largest tribal reservation in the United States, at 11,223 square kilometers (4,333 square miles) (Arizona Department of Commerce n.d.). While the reservation is held in trust by the U.S. government, both the tribal people and the U.S. government share a mixed sovereignty (Laird 1993a). Notably, the western Sonoran Desert is the only place where three nations—the United States, Mexico, and Tohono O'odham—meet together along the entire U.S.-Mexican border (Waldman 1985, 196–197).

West of the Tohono O'odham Reservation, the U.S. government established two protected areas during the 1930s largely due to concerns over the area's dwindling wildlife. This concern dated back to before 1900, when populations of the desert bighorn sheep (*Ovis canadensis*) were known to be dwindling. Although legal protection for the species had come as early as 1893 (Carmony and Brown 1993, 193), during the 1930s sportsmen's organizations and the Boy Scouts of America launched a crusade for a national game range in the region (Felger et al. 2006). Although opposition from ranchers, miners, and Arizona's governor held up the proposal, support for the game range was strong, and on April 13, 1937, President Franklin D. Roosevelt designated 1,338 square kilometers (517 square miles) as Organ Pipe Cactus National Monument (ORPI) (Felger et al. 2006; Pearson 1998, 1). At the time of ORPI's establishment, the O'odham living in the area were "kicked out" by the U.S. National Park Service (NPS) (Williams 1994), a controversial decision that many would hold as ironic given the subsequent designation of ORPI in 1976 as a biosphere reserve.

Two years after the establishment of ORPI, Roosevelt established the Cabeza Prieta Game Range to the north and west of the national monument. To "appease his wrangling bureaucrats," he placed the Range under joint management by the Biological Survey of the U.S. Department of Agriculture (USDA) and the Grazing Service of the U.S. Department of the Interior (DOI) (Carmony and Brown 1993, 196). In 1976, the Cabeza Prieta Game Range was redesignated a national wildlife refuge (NWR) under the sole management authority of the U.S. Fish and Wildlife Service (FWS) (Schumacher c.1992). Sharing a 90-kilometer (56-mile) international border with Mexico, Cabeza Prieta is the third-largest NWR in the lower forty-eight states at 3,480 square kilometers (1,344 square miles) (U.S. Fish and Wildlife Service n.d.).

In addition to the NPS and FWS, the Department of Defense has a large presence in the region. In 1941, President Roosevelt withdrew a vast tract of land to the north and west of Cabeza Prieta NWR as the Luke Gunnery Range for military training purposes (Ripley et al. 2000). After several name changes, including "Luke Air Force Range" from 1963 to 1987, Congress renamed the area as the Barry M. Goldwater Range (BMGR) in 1987 (Luke Air Force Base n.d.). After various expansions between 1941 and 1999, the Range was subsequently reduced to its current total of 7,017 square kilometers (2,709 square miles)—an area that is still large enough for more than fifty aircraft to be operating simultaneously under multiple independent training operations (Felger et al. 2006). Despite the military's active use of the area, actual damage to the vast majority of the land surface is minimal—so much so that conservationists have described the Range as a "de facto" protected area (Felger et al. 2006).

On the Mexican side of the border lie two vast tracts of land designated under Mexican law as "biosphere reserves": the Pinacate y Gran Desierto de Altar Biosphere Reserve (7,147 square kilometers; 2,759 square miles) and the Alto Golfo de California y Delta del Rio Colorado Biosphere Reserve (9,423 square kilometers; 3,637 square miles) (Felger et al. 2006). The concurrent designation of these biosphere reserves was a significant episode in the story of ISDA, in particular the saga of the Pinacate, which is described in detail further on. Other Mexican lands in the western Sonoran Desert are composed of communally owned and farmed lands called *ejidos* and small- to medium-sized communities, including the border town of Sonoyta and the beachside tourism destination of Puerto Peñasco (Pearson 1998).

In total, U.S. and Mexican federal agencies manage three extensive protected areas, two biosphere reserves, and one "de facto" reserve that cover 30,413 square kilometers (11,743 square miles), "making them the largest zone of contiguous protected desert anywhere in the Americas" (Felger et al. 2006).

MAB in the U.S. and Designation of ORPI as a Biosphere Reserve

The U.S. National Committee for the MAB Program was established in 1974 under the auspices of the U.S. Department of State. Representatives on the committee came from twelve federal agencies, as well as the National Science Foundation and the Smithsonian Institute (U.S. MAB Program n.d.). After a 1974 MAB conference in the United States and a U.S.-USSR agreement on joint biosphere reserve designation, twenty U.S. biosphere reserves were designated (UNESCO 1974). Today, the U.S. MAB Program recognizes forty-seven biosphere reserves in the country (for a chronology of the U.S. MAB Program, see Gilbert 2004; UNESCO 2000).

The U.S. MAB Program has faced two major challenges. The first has been the lack of recognition of its very existence—both from the general public and even from its own implementing government agencies. Although the NPS "informally adopted the concept as a guidepost to regional planning," Grumbine (1992, 157–158) pointed out that:

> neighboring national forests have been reluctant to participate, Congress has provided little funding for the program, and the public is hardly even aware that biosphere reserves exist. Overall, this first model of management beyond administrative boundaries suffers from nebulous goals that offer something for everybody. The hard questions are left unanswered. Biosphere reserves may yet have a future, but they are not the panacea that some wish them to be.

A former MAB coordinator for the NPS offered an even less optimistic assessment, noting that "park managers have tended to see the designation as a gratuitous honor, rather than an opportunity to [solve] management problems or strengthen bioregional cooperation" (Tangley 1988, 150). The lack of recognition was only compounded by the Reagan administration's 1984 decision to withdraw from UNESCO, a decision based on what the administration perceived as mismanagement and an anti-American agenda in several UNESCO programs (Coate 1988; Preston et al. 1989). Although the MAB Program apparently played no role in the decision to withdraw —notably, the decision did not suspend U.S. participation in either the MAB Program or several other science-based projects within UNESCO— it did indicate a low level of official support for the U.S. MAB Program (UNA-USA 1993).

The second challenge resulted from a widespread assertion by politically conservative forces in the United States that the MAB Program could lead to loss of U.S. sovereignty over private and public lands—an erroneous belief that nonetheless convinced some of impending UN plans to "enter a country for the purpose of controlling biosphere reserves" (Helvarg 1994, 225; see also Soles 1998). As will be described, such fears would become a major impediment to ISDA and the effort to establish an international Sonoran Desert biosphere reserve.

In 1976, ORPI was among the first designated biosphere reserves in the United States and the world. ORPI officials have argued that the "MAB concept has been fully realized at Organ Pipe Cactus [National Monument] with a nationally renowned resource management and research program, and a well-recognized international affairs program" (Pearson 1998, 1). Others have countered that the biosphere reserve designation was pasted on top of ORPI and that it has "never been able to fully function as a UNESCO-style biosphere reserve with designated protected [core] areas, managed use [buffer] areas and surrounding zones of cooperation" (Murrieta-Saldivar 1998, 65; Nabhan 1996).

MAB in Mexico

In the early 1970s, many Mexican conservationists saw biosphere reserves as offering "a preferable alternative" to the country's traditional system of national parks. Entomologist Gonzalo Halffter was a prominent critic of the U.S. national park model who argued that national parks could work only in countries that faced few demographic pressures, that could afford taking areas out of production, and that had both the administrative capacity and the tradition of conservation to protect the land—none of which applied to Mexico. Halffter, whom Ezcurra described as the first person to bring the biosphere reserve concept to Mexico, "rued" the fact that Mexico's national parks had never protected a representative suite of ecosystems, had not led to the advancement of ecological knowledge, and had "failed to address the social needs of the local population" (Simonian 1995, 160). As Halffter would write for the ten-year anniversary of MAB, "Rational land use and the conservation of fauna and flora have developed in parallel and do not always involve the policies of national parks. This point must be stressed, because the union of conservation and rational land use is the best hope for species and ecosystem conservation" (Halffter 1984, 429).

In 1974, a concurrent meeting in Mexico City between the MAB Program and the Latin American Zoological Conference promoted the biosphere reserve concept to biologists and conservationists (Simonian 1995, 161). In the same year and city, Halffter established the research-based Instituto de Ecología, which quickly adopted the "radically new" biosphere reserve concept by taking the lead role in designating the Mapimí and Michilía Biosphere Reserves in the state of Durango (Ezcurra et al. 1999; Simonian 1995, 161). Both of these biosphere reserves were officially recognized by the MAB Program in 1977 and by the Mexican government in 1979 (Reyes-Castillo n.d.; UNESCO 2005).

The "auspicious beginning" of Mexico's biosphere reserve program rested on the combination of Halffter's organizational skills and political support from the Durango governor and government (Simonian 1995, 162). The subsequent recognition of biosphere reserves in Mexico came from their apparent success, particularly in the Mapimí Biosphere Reserve, which reportedly drew in local people through its projects on range improvement and alternative development (Gregg 1991). Subsequently, other biosphere reserves were formally recognized by both the national government and the international MAB Program.

The legal history of biosphere reserves in Mexico is somewhat opaque. Apparently, in 1977, Mexico established the biosphere reserve as a specific type of land management unit (Condo and Bishop 1995, 40; Gómez-Pompa and Kaus 1998). However, the legal designation of "biosphere reserve" is more often attributed to the 1988 Ecological Balance and Environmental Protection Act, which sets standards for establishment and status of a Mexican biosphere reserve similar to those set by UNESCO (Laird 1993j). While Mexican biosphere reserves have "considerable administrative independence," each reserve operates under a federal or state-affiliated research institution that provides continuity to the program (Gregg 1991, 289). Mexico is one of the few countries legally to recognize biosphere reserves as an official type of land management unit (others include China and India) (Batisse 1997; de Klemm and Shine 1993; Gregg 1991, 289).

Today there are twenty-four nationally designated biosphere reserves in Mexico, sixteen of which are recognized by the international MAB Program (UNESCO 2005; United Nations Environment Program and World Conservation Monitoring Centre 2005). According to Gregg (1991, 288–289), Mexico's implementation of the biosphere reserve concept has been "exceptional," partly due to the "relative weakness of national park and other protected area systems, and the need to build centers to demonstrate ways to integrate conservation and development." Recalling what she had encountered in her extensive work with Mexican advocates and officials, ISDA organizer Wendy Laird told me that biosphere reserves—particularly in contrast to national parks—have more clout with Mexicans because they entail greater access to technological and scientific resources. However, there exists only limited evidence that biosphere reserve status has improved conservation in Mexico (e.g., Halffter 1985). Furthermore, in terms of public recognition it appears that the Mexican public generally does not understand the biosphere reserve concept (Vidal et al. 2004), and even attentive U.S. observers of Mexican culture still confuse biosphere reserves with national parks (e.g., Fisher 1994; e.g., Williams 1994; Yetman 1996, 195). Overall, Simonian (1995, 165) is probably on target in asserting that the "program as a whole can be deemed neither a success nor a failure since each reserve has produced different results."

Biosphere Reserve Planning for the Pinacate—and Beyond

In the northwest corner of the state of Sonora, the *Sierra del Pinacate* stands sentinel over that thin strip of Mexican territory connecting the mainland to Baja California. Many O'odham peoples regard the range as sacred ground, and one subgroup—the Hia Ced O'odham—regards the area as its "aboriginal homeland" and the "the home of I'thi [Itoi], the one who accomplished their creation" (Joquin 1988, 13). Geologically and archaeologically unique in North America "and perhaps in the New World," the Pinacate region is composed of a shield volcano and extensive lava fields that cover approximately 1,500 square kilometers (585 square miles; Hayden 1998, 12–13), the northern portion of which reaches across the international border into Cabeza Prieta NWR. While the Pinacate lava fields lie within the region described as the Gran Desierto de Altar, this latter name is often used to describe the dunes to the west of the Pinacate that form the "largest sand sea" in North America (404,858 hectares, or 999,999 acres) (Ezcurra et al. 1982, 68; Felger et al. 2006; Hayden 1988, 52; Hayden 1998, 14; Pearson 1998, 7; Phillips and Comus 2000, Plate 29; Yetman 1996, 196). The Gran Desierto de Altar is possibly the driest region in Mexico, with temperatures exceeding 50°C (122°F) in the summer (Ezcurra et al. 1982, 68; Yetman 1996, 196).

Considered a "world class resource" (Laird 1993b), the Pinacate forms "one of the most diverse, beautiful and representative zones of the Sonoran Desert" (Ezcurra et al. 1982, 68 [translation from original]). Whether in spite of or because of the harsh, unforgiving character of the land, scientists and conservationists often rely on superlatives to describe the area. Michael Soulé, a cofounder of the science of conservation biology, described the Pinacate as "one of the most pristine areas in the world because it's so dry" (Williams 1994). In addition, the Pinacate's "sandrock environment creates a multitude of unique habitats with tremendous biodiversity," including over 560 vascular plant species, 40 mammal species, 200 bird species, and 40 reptile and amphibian species (Walker n.d.). It has been described as "a fragile area, important for its unique species, but easily impacted by urban development. The most serious threats . . . include removal of volcanic ash, industrial and highway construction, poaching, agriculture, and water extraction" (Laird 1993b).

The Pinacate is also important archaeologically because it attracted human occupation from the surrounding dry desert with its rain-fed tinajas, or natural water tanks, and because ancient human remains there are extremely well preserved (Hayden 1998, 27). The Pinacate has thus fostered a "tradition of rogue scholars," including some who have long argued that humankind entered the North American continent up to 40,000 years ago (Hayden 1998, 27). As the mythological birthplace of the radical conservation group Earth First! (Manes 1990, 4; Parfit 1990; Williams 1994; Zakin 1993, Chapter 7), "rogue" conservationists have reportedly found inspiration in the Pinacate—although Dave Foreman, one of the principal founders of Earth First!, has averred to me that this is indeed pure myth. Nonetheless the Pinacate's grand aura is well conveyed in an aphorism attributed to conservationist and author Edward Abbey: "Saguaro National Monument is high school, Organ Pipe Cactus National Monument is undergraduate, Cabeza Prieta is graduate school, and Pinacate is post doc, so far as the Sonoran desert goes" (Zakin 1993, 116).

As early as 1943, the Mexican government investigated the possibility of establishing a game refuge in the Pinacate and collaborated with the United States in a small, research-oriented transborder conservation initiative (Pearson 1998, 6). Secretary of the Interior Stewart Udall recalled that when he was investigating the possibility of a Sonoran Desert National Park during the 1960s,

> I know we talked about the Pinacate as a remarkable geographic area in Mexico. I think there were preliminary talks of an international park then, not so much with government officials but by conservationists in Sonora. But I have to say, in all honesty, the government of Mexico at that time didn't seem to have a very high priority of having national parks jointly with the United States on the border. . . . The idea of an international park, the U.S. and Mexico, was kind of a dream at that point. We didn't get down to practical cases. (Udall 1997, 316–317)

Pearson (1998, 7) notes that in 1975 "there was talk of establishing international formal working agreements with Mexico," followed the next year by various cooperative activities across the border.

In March 1979, the Mexican government designated 28,660 hectares (70,790 acres) of the *Sierra del Pinacate* as a Protected Forest Zone and Wildlife Refuge under the director of reserves and recreation areas of the Ministry of Agriculture and Water Resources (SAHR) (Búrquez 1998, 74; Búrquez and Castillo 1993; Hayden 1988, 52; Pearson 1998, 7). While Ezcurra noted that the legislation establishing the refuge's boundaries was so vague as to mean almost nothing, he added that the designation of the Pinacate as a protected zone did at least indicate "the intention on the part of the federal government to protect this area."

In addition to these initiatives at the federal level, the state of Sonora was also actively investigating the possibility of biosphere reserve status for the Pinacate. In 1980, Dr. Samuel Ocaña, a former prominent surgeon and "devoted conservationist," became the governor of the state of Sonora (Ezcurra et al. 2002; Yetman 1996, 119). In 1984, he was instrumental in establishing the Centro Ecológico de Sonora as a government-funded research and environmental education institute based in the state's capital city, Hermosillo. With support from The Nature Conservancy, the Centro Ecológico had been working on a "vast ecological plan" for the whole state of Sonora, and both the Pinacate and the Upper Gulf were included in that process (Steinhart and Blake 1994, 145). As part of the process, Ocaña requested Gonzalo Halffter and the Instituto de Ecologóa to conduct scientific investigations into the Pinacate as a possible biosphere reserve (Búrquez 1998, 74; Condo and Bishop 1995; Halffter and Ezcurra 1982).

Halffter sent biologist Ezcurra, beginning a three-year scientific investigation that resulted in a "master plan" for a recommended biosphere reserve (Búrquez 1998, 74). Looking back at the long process of designating El Pinacate as a biosphere reserve, Ezcurra later reminisced that "the whole thing probably started in 1980 when I first went in an old jeep into the Pinacate and met people like Carlos Nagel or Gary Nabhan or Sylvester Listo"—three individuals who would be involved in ISDA to varying degrees.

Hopes for designating the Pinacate as a biosphere reserve seemed high at this point, high enough for Halffter to report to a 1981 international conference, the VI Symposium on the Gulf of California Environment, that "El Pinacate (Sonora) is in the last phase of setting up a reserve of half a million hectares" (Halffter 1984, 430). The following year, a binational workshop of scientists and managers met "to discuss and recommend the implementation of a Sonoran Desert Biosphere Reserve covering areas in both countries," and it was during that time that the on-site manager of the Pinacate reserve cooperated with ORPI on several initiatives (Anonymous 1994a; Condo and Bishop 1995; Pearson 1998, 7). Also, in 1982, the group Amigos de Pinacate, part of the U.S.-based conservation organization Friends of ProNatura (FPN), sponsored several environmental initiatives in the Pinacate region (ProNatura is short for Asociación Mexicana Proconservacion de la Naturaleza, a prominent Mexican conservation organization founded that now primarily works out of several regional offices; see Ezcurra et al. 1999, 176) (Pearson 1998, 8). Despite these myriad activities, the effort to designate the Pinacate as a biosphere reserve stalled. In 1982, before the biosphere reserve assessment for the Pinacate process was completed, Mexico's protected area system was restructured and the Pinacate was designated an "ecological reserve" under dual administration by SAHR and the Ministry of Urban Development and Ecology (SEDUE) (Búrquez 1998, 74–75). Although the two agencies signed a cooperative agreement in 1984 to protect and manage the area cooperatively, the efforts reportedly "failed to yield satisfactory results." In all likelihood, the bureaucratic wrangling between these two agencies was, to say the least, not conducive to biosphere reserve thinking (Búrquez 1998, 75; Pearson 1998, 8). Furthermore, as Ezcurra explained to me, the Mexican federal government

> was at that time opposed to decreeing protected areas along the Mexico-U.S. border. And this was a strategic thing; they were opposing that because, since the early thirties of this century, the American government has been talking about making parallel protected areas in Mexico and the U.S. on some critical sites along the border. And Mexico perceived that the setup of national parks along the Mexico-U.S. border, like Big Bend or Organ Pipe, were really things that the U.S. did to define its boundaries and territories and to have control of the border—to keep the border under some sort of governmental control. So the Mexican government at that time was not really too sympathetic towards the idea of a protected area along the border, and the whole project of the Pinacate never took off.

Despite the recalcitrance of the Mexican federal government, several agency representatives participated in convening a 1988 Symposium on the Pinacate Ecological Area, the general purpose of which was "to identify mutual needs and interests compatible with the biosphere reserve concept" for the region (Condo and Bishop 1995, 40; Environment Committee of the Arizona-Mexico Commission 1988; Keystone Center 1996; Laird et al. 1997, 309–310). Hosted by the Centro Ecológico de Sonora, the conference brought together scientists and conservationists from both sides of the border. Significantly, the 1988 symposium was the first occasion of both the Tohono and the Hia-Ced O'odham being included in such a transborder forum (Laird et al. 1997; Sonoran Institute and International Sonoran Desert Alliance n.d.). To understand why that event was so significant—and so long in coming—it is critical first to understand the history of the region's indigenous inhabitants.

Traditional archaeological dogma long held that humans first crossed into the North American continent approximately 11,000 years ago. That hypothesis has been challenged on many fronts, however, and most archaeologists would now agree that the confirmed date of the earliest arrival was at least 5,000 years earlier (Page 2003). One of the early proponents of an earlier arrival date was Julian Hayden, a colorful personality widely celebrated in the archaeological and natural history literature of the Sonoran Desert. Based on both his extensive studies in the Southwest region and his intensive examination of the Pinacate region,

Hayden concluded that people of the San Dieguito Complex inhabited the Adair Bay region from approximately 37,000 to 9,000 years ago, and that "Indians have lived in and about the Pinacate for perhaps far more than 40,000 years" (Hayden 1998, 25, 27).

Whatever position one takes on the unresolved puzzle of when humans first entered the North American continent, archaeologists generally agree that prehistorical inhabitants of the western Sonoran Desert had a significant effect in shaping the landscape (Sheridan 2000, 106). According to Hayden, after the San Dieguito Complex and subsequent Amargos Complex (7,000–1000 BC) came the Hohokam, contemporaries of the more widely recognized Anasazi (more respectfully referred to as the Ancestral or Ancient Puebloans) (Hayden 1998; Waldman 1985, 5). Inhabiting the area from the east coast of the Gulf of California northeastward to an area bounded approximately by an arc formed by the present-day sites of Nogales, Tucson, Phoenix, Gila Bend, and Ajo, the Hohokam civilization was marked by pit houses, pottery, and the largest system of irrigation canals in pre-Columbian North America, constructed between 300 and 1450 AD (Krech 1999, 45; Sheridan 2000; Vesilind 1994, 44; Waldman 1985, 16–17). The Hohokam culture dispersed around 1500 AD for a number of possible reasons, including climate change and disease (Sheridan 2000).

Archaeologists have not settled a debate over whether the contemporary O'odham peoples occupied the land previous to, contemporary with, or after the Hohokam (Erickson 1994, 16; Sheridan and Parezo 1996, 116–117). Whatever the case, the O'odham have been the predominant indigenous group out of a minimum of thirteen that have occupied or acquired resources from the area since the advent of European exploration in the Sonoran Desert (Emanuel 2000, 11). Approximately 18,000–21,000 O'odham still speak the native language and live in south central Arizona, while another 1,000–2,000 live across the border in the adjacent state of Sonora (Nabhan 2000). The O'odham comprise several subgroups, the largest of which is the Tohono (or "desert") O'odham, once commonly—and unfortunately sometimes still—referred to as the Papago or "bean people." The Tohono O'odham Nation has approximately 25,000 enrolled members, although only about 11,000 live on the reservation (Tohono O'odham Community College n.d.). A smaller group known as the Hia-Ced O'odham, or "Sand Papago," once lived in the land west of the Tohono O'odham reservation, much of which is at present occupied by larger blocks of U.S. and Mexican federal lands (Nabhan et al. 1989). The transition zone between the Tohono O'odham and the Hia-Ced O'odham traditionally ran northwest-southeast near the sites of Quitobaquito and Quitovac (Nabhan et al. 1989). Although some historical accounts describe the tribe as extinct, over 1,000 individuals today describe themselves as Hia-Ced O'odham (Nabhan et al. 1989; Sonoran Institute 1999; Williams 1994). Lorraine Eiler (1992, 3), a Hia-Ced O'odham and one of the early active participants in ISDA, has argued that despite being allowed to enroll in the Papago tribe in 1982, the Hia-Ced have "never been properly acknowledged as Indian People" by the U.S. government.

The O'odham face problems similar to those confronting other Indian tribes in the United States, including unemployment of over 50 percent (Sonoran Institute 1999b), the loss and degradation of cultural heritage (McGuire 1995), alcoholism, and the deterioration of health due to their drastic shift in diet with the adoption of a "European" diet (Nabhan 2002). Two of the most severe impacts of European culture have been more localized. The first is in regard to the desert's most precious resource: water. Whereas O'odham lifeways revolved around capturing the Sonoran Desert's biseasonal rainfall, the Caucasian newcomers "mined" water from the ground. Their wells, as the noted regional author Charles Bowden described the impact, helped transform the O'odham "from a people who had lived off the land to a people who happened to live on the land" (Bowden 1977, 7).

The second impact was the establishment of the international border, which split the O'odham's homeland in two. The O'odham have long protested the resultant cultural and linguistic effects and even garnered sufficient federal-level attention to establish a "Tohono O'odham in Mexico" office (Rodriguez et al. 1988). However, the problem has never been sufficiently resolved, and indeed many O'odham came to participate in the International Sonoran Desert Alliance primarily in the hope of resolving their frustrations over the border.

Because of the Pinacate's "critical importance" to the Hia-Ced's culture, history, and identity, the O'odham argued at the 1988 symposium that the concept of designating the Pinacate as an international biosphere reserve was "of paramount interest" to the Tohono O'odham Nation (Joquin 1988, 13). In essence, the Hia-Ced were concerned that "their aboriginal rights have been violated and may continue to be violated, or at best, not even considered, by the framers of such an international biosphere designation" (Joquin 1988, 13). The O'odham then recommended several steps for incorporating the Hia-Ced into planning and management for the Pinacate, and "requested the represented entities of the Pinacate Symposium to recognize the Hia-Ced O'odham as the original inhabitants" of the area (Eiler 1992; Joquin 1988, 13).

The O'odham's presence and statement at the symposium marked a turning point in both land management and land management advocacy in the Sonoran Desert. Although it would often be little more than lip service, from that point on it became increasingly *de rigueur* for both conservationists and government officials at least to cite the O'odham's traditional connections to the land. And so when a group of participants at the 1988 symposium agreed that there was a need for "a larger public forum . . . to promote dialogue among residents of the Sonoran Desert," it was clear to them that this new initiative could succeed only with extensive participation by the O'odham (Laird et al. 1997, 310).

ISDA participants have credited the 1988 research symposium as the most significant precursor leading to the establishment of ISDA five years later. Carlos Nagel, one of the principal organizers of both the symposium and ISDA, later referred to the former as "the bioregional process that was initiated in 1988" (Nagel 1992a). He also noted that he "had not realized the extraordinary effect of the 1988 Pinacate Conference. . . . I believe the stage is set for an organization, with the skills that need to be applied to further land use planning" (Nagel 1992b). Both as the president of Friends of ProNatura and as a consultant "in cross-cultural facilitation and translation" between Hispanic and Anglo cultures (Bennett 1989, iii), Nagel was in a key position at the time of the symposium—particularly inasmuch as he had also been contracted by the National Park Service to draft *A Report on Treaties, Agreements, and Accords Affecting Natural Resource Management at Organ Pipe Cactus National Monument*. The report made several references to the MAB Program and the possibility of an "International Biosphere Reserve," emphasizing that it would be "especially appropriate to incorporate more traditional people and land use in and near the Monument" (Nagel 1988, 11). Most significantly, the report urged "that mechanisms be developed to encourage professionals in the various cultures to meet face to face and to know each other" (Nagel 1988, 12). Nagel's aspirations for such a mechanism were to be realized, but it would take a half decade and the appearance of a nontraditional institutional partner on the Sonoran conservation landscape.

The Sonoran Institute and the 1992 Border Land Use Forum

After the 1988 symposium, Nagel was floating the idea of "a binational town-hall meeting" to promote international cooperation in the western Sonoran Desert (Laird 1994b, 10). Yet while both the MAB Program and ProNatura supported Nagel's documentation of a "widespread interest in assessing the feasibility of a cooperative regional program that would empower a wide range of stakeholders in implementing biosphere reserve concepts," neither was able to provide financial support (Gregg 1992). Nagel was apparently stymied—and might have remained so had it not been for the arrival on the scene of Luther Propst, an innovative conservation thinker, and his hybrid conservation and community development organization with a name eponymous to the region: the Sonoran Institute (SI).

Three years as a lawyer representing land owners and environmentalists had left Propst disillusioned. "I became rather skeptical about this field of legal practice," he later wrote, "because as I had the opportunity to travel around the country, it became clear to me that the places that were really solving land use problems were places that generally avoided polarized rhetoric and litigation" (Propst 2003, 1). Subsequently, working with the Washington, D.C.–based World Wildlife Fund (WWF) in a program entitled Successful Communities, he was inculcated in the organization's "basic approach to conservation" in developing countries, which consisted of building "productive, mutually beneficial relationships between managers of protected areas and the people that live nearby" (Propst 2003, 1). "You can't have effective place-based conservation," he told me, "if you rely entirely upon the public land managers." Propst brought this philosophy to Tucson and, with critical yet minimal startup funding from WWF-U.S., established SI in Tucson in 1990.

SI eventually became a well-respected organization in the North American conservation world, and Propst came to be described as "one of the nation's leading proponents of local efforts to build community and enhance both conservation opportunities and the quality of life in small towns" (Chronicle of Community 1997). (Serendipitously, in 1997 SI opened an office in Bozeman, Montana, where its staff have been actively involved in Y2Y; see Chapter 4.) However, its first years were marked by both financial struggle and controversy over SI's connection to another of Propst's projects established at the same time—the Rincon Institute, the mission of which was to protect Tucson's Saguaro National Park (east unit) and surrounding lands. To fund the Rincon Institute, Propst's approach was to bypass "traditional sources of revenue" and instead "go after the funding that becomes available through the development process" (Chronicle of Community 1997). In other words, he argued, when development occurs—and it will occur—the developers should make a concomitant payment toward conservation. The catch was that Propst set up the Rincon Institute with funding from Donald Diamond, a Tucson developer notorious to local conservation groups. Furthermore, the Rincon Institute existed only in an advisory capacity. Wendy Laird, who was one of SI's first employees and later became one of the principal organizers of ISDA, told me she recalled "signs on street corners that said, 'Don Diamond doesn't own this town'—but the reality is, yes, he does." Laird noted that from the moment she began working in the western Sonoran Desert, much of her job entailed differentiating SI from the Rincon Institute, as well as defending SI from accusations of being used as a "green" shield by wealthy developers.

Indeed, working as a volunteer for her first nine months at SI, Laird was frustrated by accusations of SI having been bought out. Despite the initial lack of remuneration, Laird proved to be a highly motivated worker who also had the skill of motivating others. While Propst focused his efforts in the broader Tucson region, she began putting in eighty-hour work weeks in the western Sonoran Desert town of Ajo. SI's cooperative approach meshed well with that of Nagel (who later joined the board of the Sonoran Institute), and before the year was out they had partnered to bring together different actors in the Sonoran Desert for a "regional consultative forum" to enable all potential stakeholders to "define the geographic area for cooperation and to develop a coordinating structure for maintaining future dialogue" (Condo and Bishop 1995, 43).

At the time, there appears to have been significant interest in and support for a conference on land use in the western Sonoran Desert. A few months before the conference, Laird presented the idea of a "proposed regional forum on land-use changes in the western Sonoran Desert border region" to the Western Pima County Community Council (WPCCC) and the Ajo District Chamber of Commerce. According to the local paper, her work produced an "outpouring of traditional local views on the government handling of federal lands" (*Ajo Copper News* 1992). In addition, William P. Gregg Jr., the U.S. MAB program coordinator with the NPS, wrote that the proposed "timely" forum could invigorate "regional interest in the biosphere reserve concept" and should help identify "important opportunities to improve cooperation in developing the knowledge, skills, and attitudes needed to achieve ecosystem sustainability" (Gregg 1992).

Yet even as there seemed to be high interest in holding the event, few were initially interested in supporting it financially. Laird recalled that she spent long hours raising funds and finding sponsors, garnering nearly thirty of them by the time the forum took place in Ajo on October 9–10, 1992. The most significant partner, recalled Propst, was the Lincoln Institute of Land Policy, a Massachusetts-based think tank that offered critical logistical support and made a sizable grant to support the conference. Laird also saw that one of the most important sponsors in terms of "obtaining legitimacy in the region" would be The Nature Conservancy (TNC). She went to talk with TNC's director for Northwest Mexico, Susan Anderson, at which point Anderson made a key decision: TNC would support the conference on the condition that Laird could ensure the participation of Exequiel Ezcurra, then Mexico's Secretary of the Instituto Nacional de Ecología. Laird called Ezcurra out of the blue, and, soon after, he committed to come.

Entitled "Land Use Changes in the Western Sonoran Desert Border Area: A Regional Forum," the conference became generally referred to as the 1992 Land Use Forum. Over two hundred individual participants included citizens of all the three nationalities; representatives from NGOs, Chambers of Commerce, and Native American organizations; and federal, state, and county government officials (Laird 1994b; Laird and Anderson 1996a; SI and FPN c.1993). The large showing was due at least partly to the extensive preconference publicity for the event, the bulk of it having been shouldered on Laird and Nagel.

Various speakers and panels at the forum covered a wide range of resource and cultural aspects of the region, including free trade, community development, growth and land use change, natural resource use, and O'odham history and culture. Ethnobotanist and prolific author Gary Nabhan promoted the biosphere reserve concept for the region. A panel on public land issues convened government officials from ORPI, Cabeza Prieta NWR, the Bureau of Land Management (BLM), the Department of Defense (DOD), and Parque Natural del Pinacate y del Gran Desierto (Lincoln Institute of Land Policy and Sonoran Institute 1992). The last day of the conference included a science panel covering "binational efforts" in archaeology, botany, and biodiversity protection (specifically focusing on bats and Sonoran pronghorn) (Laird n.d.; Lincoln Institute of Land Policy and Sonoran Institute 1992). Also on the last day of the conference, there were several small workgroup sessions intended (1) "to develop a statement of values, concerns, projected conditions, and possible tangible steps to guide future action in the region by private citizens, businesses, and state, local, and federal government agencies in the region" and (2) to encourage "informed communication and interaction between diverse people who rarely have the opportunity to talk with one another" (Sonoran Institute 1992, 1). The work group sessions identified several challenges facing the region, ranging from the practical (e.g., NAFTA and the future of Highway 85, a *major* throughway of the region) to the intangible ("maintaining economic vitality" and "building better multi-cultural understanding") (Sonoran Institute 1992, 1).

As with the 1988 symposium, Laird recalled that one of the conference's most important achievements was the engagement of representatives from the Tohono O'odham Nation, including the tribal chairman Josiah Moore, who, as a vocal proponent of asserting the tribe's identity, had been a key figure in the reservation's formal name change from "Papago" to "Tohono O'odham." In addition to the Tohono O'odham, the less recognized Hia-Ced O'odham made a group statement on their cultural legacy in the region. "As Hia-Ced O'odham (Sand Papago), we are the original people of this territory," began the statement, which went on to describe how important "resources, sacred sites and burial grounds" throughout the area had become "inaccessible to the Hia-Ced O'odham since the establishment of the International Border, National Parks, and Federal Reserves." The statement then made the following request: "Since 1980, we have been pursuing a small part of our aboriginal lands to become the twelfth district of the Tohono O'odham Nation. The area identified is 106,680 acres of BLM land in the Ajo Valley, between Ajo and Why, Arizona. We want this land so we can have a place to call home" (Eiler 1992, 1–3).

Although the request "certainly dropped some people's jaws" and generated concern over potential "negative reaction from the folks in Ajo" (Broyles 1992), the Hia-Ced O'odham statement included the conciliatory language that "through this gathering we hope to establish and develop a network with all entities in this area.... We want to see the planning for economic growth include the preservation of natural resources and respect of the elements essential for our survival" (Eiler 1992, 4). It also noted that the Hia-Ced "looked forward" to good relations with the

Mexican government in the planning and administration of the proposed Pinacate MAB project (Eiler 1992, 3). This desire to connect with other cultures in the Sonoran Desert would become a critical component in the establishment of ISDA.

Overall, the conference appeared to be well received. There were, of course, detractors. At least one prominent conservationist argued that the conference did not adequately address many of the looming land management issues facing the region (Broyles 1992, 6), while a businessman from the border town of Lukeville expressed concern over ulterior motives of at least one government agency, as well as the lack of any official speakers from the business sector (Gay 1992). Despite those criticisms, Laird (1993e) argued that the conference was "the first time Ajo had a regional conference of that size, the first binational, multi-cultural event of that size in the area, the first time scientists, academics, land managers, residents, and business leaders had met each other, the first time we agreed about the importance of protecting the region (including the region's communities) and working together as a diverse group to strengthen relationships across borders."

In short, she observed, "We began to cooperate." And thus was born the partnership that "cemented the foundation for the alliance" (ISDA 1993a; Laird 1994b). It is worth noting that documents from this time period indicate that Laird's leadership was critical to sustaining the effort to build ISDA. Broyles (1992) noted that at the Land Use Forum, she "did a remarkable job of pulling together so many speakers, many of whom are a bit cantankerous and headstrong. Her leadership, organization, and diplomacy were remarkable." Over the course of the next two years, Laird would become the person most intimately involved in ISDA's development.

The Faces of ISDA

As the preceding section has shown, ISDA was born out of both the broad intellectual heritage of the MAB Program and the specific effort to establish the Pinacate as a biosphere reserve. A key historical question is the degree to which ISDA subsequently tied itself back to both the MAB program and biosphere reserve concepts. More specifically: to what degree did ISDA internalize biosphere reserve concepts and promote biosphere reserve designations in the Sonoran Desert?

The answer to that question is complex, and as many ISDA participants pointed out to me, the question does not necessarily steer toward a comprehensive understanding of ISDA's multifaceted identity. For despite its roots in biosphere reserve concepts, much of what occurred under the aegis of ISDA had little to do with the land management issues associated with biosphere reserves. Thus, it is important to recognize that *one* of ISDA's identities was as a regional forum that consciously did not revolve around the biosphere reserve ideal. Understanding how ISDA took on that identity, as well as its multiple other identities, requires some explanation.

Going into the 1992 Land Use Forum, the organizers clearly had biosphere reserves in mind as a major theme for the conference. Laird and Nagel, with Propst offering support in the background of Tucson, were all versant in biosphere reserve concepts, as well as passionate about using them to protect the biological and cultural heritage of the Sonoran Desert. Yet Laird later recalled that despite the MAB Program's origins in tying nature conservation together with human development—and the emphasis it had placed in the early 1980s on working with local stakeholders (see Chapter 2)—nobody had figured out how to do those things in practice. "U.S. MAB and UNESCO don't get into those details," she said.

Consequently, both at the forum and afterward, Laird and Nagel understood that ISDA could not be some sort of simplistic biosphere functionary and by no means did they limit the discussion to core areas, buffers zones, and the like. Rather, they shared a strong belief that public participation would reveal a widespread desire to protect the desert ecosystems and create sustainable economies. And as pointed out in Chapter 2, that approach was becoming increasingly institutionalized within the international MAB Program itself. Accordingly, they acted on the assumption that "actually motivating and involving community residents" was a fundamentally important aspect of any approach to environmental management (Laird and Anderson 1996a) and that "linking key stakeholders (i.e., community residents, government officials, NGOs, etc.) and resolving critical issues collaboratively will result in better, more enduring results than traditional confrontation and polarization that often occur when dealing with environmental concerns" (Laird 1995).

In sum, long before the Land Use Forum, it was clear to ISDA's leaders that generating local participation would require coverage of a much broader suite of issues that ranged outside—sometimes far outside—those traditionally addressed under the aegis of "biosphere reserve management" (SI and FPN c.1993). Their open-minded approach was only reinforced at the forum, where they undoubtedly noted tension over the degree to which different participants wanted to focus on biosphere reserve concepts. Indeed, many individuals who participated in ISDA did so because they were personally concerned with issues such as healthcare, indigenous rights, and border crossings. Due to the fact that some of them had "very basic needs," as one early participant noted, they often simply did not understand the thinking behind biosphere reserves.

Over time, ISDA's leadership would gradually steer ISDA more directly toward the target of an international biosphere reserve for the Sonoran Desert. To confound matters from a historical perspective, ISDA also had an indirect role in how biosphere reserve concepts were being applied in a domestic context within Mexico. Thus, the story of ISDA is composed of three principal threads: (1) the establishment and development of ISDA as an independent organization, (2) ISDA's role in the designation of the Pinacate as a biosphere reserve, and (3) ISDA as one factor in the effort to create international biosphere reserves in the Sonoran Desert. It was a complex scenario, and it was not surprising to find during my interviews that only a few ISDA participants from those early years made a clear distinction between ISDA and biosphere reserve concepts. Having devoted so much time and energy to the initiative, Laird and Nagel perhaps best summarized the relationship between the two. Describing the biosphere reserve perspective as "permeating" what ISDA does, Nagel characterized the two as "parallel" phenomena. And while Laird at first described the goal of "developing an international biosphere cooperative" as one "extension" of ISDA (Laird 1995), she later described ISDA's formation as "concurrent" with biosphere reserve planning in the area (Laird et al. 1997, 309).

"Permeating," "parallel," "extension," "concurrent"—all of these words struggle to describe the relationship between the three dominant threads. Despite their being tightly knitted together, I will attempt to unravel these threads in the following three sections in order to clarify exactly what ISDA was able to accomplish.

Establishing ISDA as a Voice of and for the Desert

Today, drug trafficking and illegal immigration constitute perhaps the two most salient problems of the border region. Although both activities have occurred for decades, their rates grew dramatically through the 1990s—a result, to no small degree, of the tightening of the border in the binational metropolitan areas to the west and east of the western Sonoran Desert (Lieberman 2000, 997). However, it is important to keep in mind that the problem of drug trafficking was not a significant driver of either ISDA's original conception or its subsequent programmatic orientation. In terms of illegal immigration, on the other hand, ISDA was indeed born into the broader context of the border region's demographic trends, and it is important to review how these trends helped lead to the creation of ISDA.

Population issues on the border can be broken down into the two primary categories of immigration and growth. Illegal immigration from Mexico to the United States arguably sets the basic tone of the debate over how much legal immigration the United States should allow—and in so doing instills an ever present source of tension in the broader U.S.-Mexico relationship. Notably, policy debates in the United States over immigration tend to blur the distinction between the political right and left, where certain vocal arms of the environmental movement find themselves entwined with those of the nationalistic right in pushing for lower levels of legal immigration and stricter control of illegal immigration. Behind such broader policy debates lie the myriad human miseries adhering to the quest for *el Norte*, a quest that in the western Sonoran Desert repeatedly manifests itself in the desiccated bodies of countless ill-informed—often purposefully misinformed—migrant workers who set out into vast stretches of desert with nothing but cheap plastic water jugs grasped in their hands (see, for example, *Ajo Copper News* 2000, 2001; Alaimo and Barios 2000; LeDuff 2001; Zeller 2001).

While populations have shifted and grown due to both legal and illegal border crossings, each country has found its border towns and cities expanding rapidly due to domestic immigration (Walker and Pavlakovich-Kochi 2003). In Mexico, waves of migrants from the south of the country come to the state of Sonora either to make a living or to attempt to cross the border. In the single year after ISDA's conception, "1.8 million Mexicans—mostly from the states of Michoacán, Jalisco, Guanajuato, Zacatecas, and Oaxaca—moved to northern Mexico in 1994, of whom 797,000 moved on to the United States" (Oppenheimer 1996, 288). The major urban centers on the U.S. side have also seen explosive growth. As Abbott (1993, xii) describes, "It takes a vigorous act of will to remember that Phoenix in 1940 was the same size as present-day Yakima, Washington (186,000)." Comprising multiple "boombergs" such as Tempe, Glendale, and Mesa, greater Phoenix today covers an area the size of Delaware and is home to approximately 3.3 million people (CensusScope n.d.-a; Wikipedia 2004); smaller Tucson increased 26 percent between 1990 and 2000 and has a current population of over 800,000 (Census-Scope n.d.-b). Although both Phoenix and Tucson lie to the north and east of the western Sonoran Desert, population growth in those cities has resulted in increased use of its public lands—much of it for recreational activities.

From one perspective, such rapid growth and its concomitant nefarious activities arose in the context of impenetrable cultural barriers that straddle the "Tortilla Curtain" of the international border (Curtis 1992). Although it is a cliché to describe borders as nothing more than artificial human constructs, anybody who has crossed the U.S.-Mexico border knows that it is as real and undeniable—not to mention imposing and intimidating—as "artificial" can be. No doubt, it is an "abrupt demarcation" and a "real, hard, and physical fence" (Curtis 1992; Laird 1994a, 1). As Yetman summed up his three-plus decades of crossing the Sonora-Arizona border, "One thing has not changed: the dramatic difference one sees when crossing the border. The language, culture, customs, even the countryside seem to change instantly. Nowhere else on our planet is a political boundary accompanied by such a vast social difference" (Yetman 1996, 4).

Yet despite the myriad geographical and cultural differences, many inhabitants recognize the border as running through a distinct region with a shared biological and cultural heritage—a heritage leading to what Oppenheimer (1996, 287) has described as "an inevitable cultural symbiosis" between the two countries' border states (see also Allen 1997). As Curtis (1992, 376) has written about this region called both *la frontera* and MexAmerica, the border functions

> as a kind of linear third country or special domain with its own identity and character. This zone of "overlapping territoriality," as it has been labeled, has produced a hybrid culture, one that is part Mexican and part American, similar to yet different from the cultural mainstreams found in the interiors of the two nations.... It is suggested that the so-called *fronterizos*, the border people, share a common experience and are tied not only by geographical proximity but also by interdependence, mutual interests, and transborder concerns ...

In forming and leading ISDA, staff would often proffer similar reflections. Laird described the region's "border culture" as "an integrated whole;

seen by many as a separate nation state. Border communities are *not* Mexican, *not* American, but just BORDER. [They] are noteworthy for their shared cultural heritage, economic interdependence, and the relative free flow of ideas, people, goods, and services" (Laird 1994a). ISDA coordinator (and, later, executive director) Reggie Cantú described the border as "almost a third country" (Zakin 1997).

Economically, the region has suffered through the boom-and-bust cycles of an extractive mining economy—although it was mostly bust during the decade leading up to the birth of ISDA. One of the more hard-hit areas was the town of Ajo, the "heart of the region" that contains the largest U.S. population center in the western Sonoran Desert (population 4,000) and is home to ISDA's main administrative headquarters (Nagel 1992b). There, life changed dramatically when Phelps-Dodge closed the New Cornelia copper mine in 1984. The closing induced a severe localized depression that left many of the town's residents—including the O'odham, who had lived in a segregated part of the community known as Indian Village—without employment opportunities and, as Lorraine Eiler put it, "near homeless." The economic slump not only set the context for ISDA, but also disrupted the town's social structure to the point where the closing remains to this day a subject of resentment and frustration (although a shift to service industries for tourists and the steadily increasing in-migration of elderly pensioners have since helped bring about some economic growth to the region) (Kingsolver 1996; Laird and Anderson 1996a; Nagel 1992b; Vesilind 1994).

These complex cultural and economic issues set the stage for the 1992 Land Use Forum. Coming out of the forum, numerous participants, led by Laird and Nagel, sensed "the need for continuing the type of positive dialogue begun at this first forum and directing greater effort at overcoming indifference" (Sonoran Institute 1992, 2). Over the next few months, this group would endeavor to find a name for themselves, to define a mission, to incorporate, to build a board that represented the region's diverse human populations, and to find start-up funding.

Finding a Name

Less than two months after the 1992 Land Use Forum the group that was to become ISDA had their first meeting in the town of Sells on the Tohono O'odham Reservation. Through the remainder of 1992 and into 1994, a group averaging "40+ individuals per meeting" continued to attend a series of smaller follow-up "town hall" meetings in which "*anyone* can participate and *anyone* can attend" (Laird 1993g, 1994c).

From the start, the idea of forming some sort of organization—even if only a very loose-knit network—permeated the meetings. At the first meeting, the group named itself the Western Sonoran Desert Border Committee. Subsequent variants included the Regional Steering Committee, the Western Sonoran Desert Border Area Committee, the Comité Consenso, the Comité Consenso Regional, the Consenso Sonorense Regional, the Consenso Sonorense Internacional, the Sonoran Desert Alliance, and, most commonly, the Comité Consenso Internacional (CCI 1993a, 1993b; CSI 1993a, 1993b; Laird 1992, 1993a; SI and FPN c.1993).

The first documented use of the name International Sonoran Desert Alliance appears on a March 11, 1993, memo to the participants from Laird and Nagel, whose guiding hands are apparent in the plethora of memos sent out to participants during this time period. Despite concern over taking the notion of "consensus" out of the name, the group formally adopted the title of International Sonoran Desert Alliance at the end of July 1993 in order "to reflect the region, so that it could be translated into English, Spanish and O'odham, include an international component, be short, and yet retain the spirit of consensus which is what the group has been built upon" (Laird 1993i). Unfortunately, as Eiler later pointed out to me, there was no direct O'odham translation for the name. (For convenience, the rest of this chapter refers to "ISDA" and "ISDA participants" before and after formal adoption of the title.)

Defining a Mission

Enmeshed in choosing a name was defining a mission, and to some degree ISDA's tripartite mission has been immutable from its conception to the present day. Not in order of priority, the three components have been environmental protection, sustainable development, and international communication. Despite this consistency, however, a good deal of time was spent translating those broad goals into a coherent mission statement. At the first meeting after the Land Use Forum, the "overall opinion" was "to promote goals of comprehensive planning by developing a regional 'vision for the future'" and to elicit local participation in that process (Laird 1992). Subsequent iterations integrally combined *who they were* with their *mission*. For example, an early 1993 document authored by the Sonoran Institute and Friends of ProNatura described the group as "a community-based, multi-cultural initiative established to protect the diverse and unique biological resources of the western Sonoran Desert, while meeting the economic needs and aspirations of the region's residents." The document went on to say that the group forges a partnership between residents, business leaders, resource managers, and conservationists. *"No similar effort like this—involving Anglo, Hispanic, and Indigenous people—exists anywhere along the U.S.-Mexico border, despite a conviction by many that this type of ecosystem, grass-roots approach is the only one that will work"* (SI and FPN c.1993).

By mid-1993, Propst, with input from Laird and other ISDA board members, had developed the following mission statement for a draft of the "Articles of Incorporation":

> The purposes for which the corporation is organized and the character of affairs which the corporation intends to conduct are: (a) To promote environmentally sustainable and culturally sound economic development while protecting the natural and cultural heritage of the western Sonoran Desert U.S.-Mexico border region by: enhancing international communication and cooperation; fostering research, evaluation and dissemination of information about the region; and encouraging meaningful public participation in decisions affecting the region. (CSI 1993a, 1993b)

This statement remains substantively unchanged in ISDA's current bylaws, and from 1993 on, Propst's formulation would be reflected in numerous memos, letters, pamphlets, and other documents—although the emphasis would change slightly, depending on the author and the audience (see, for example, CCR c.1993b; ISDA 1993a, 1994b; Keiller 1993; Laird 1993c, 1994c; Laird and Nagel 1993c; Quintana Silver 1994a). Often, the emphasis was put on cooperation, as in the following:

> The International Sonoran Desert Alliance is a unique process that reflects the conviction that a community-based, cooperative approach which involves the people of the diverse border region of the United States, Mexico, and Native America . . . offers the best alternative for solving some of the complex economic, environmental, and cultural issues of the Sonoran Desert region. (ISDA c.1996a, c.1996b)

Most succinctly, as formally stated in its newsletter, *Vista*, ISDA's mission is "to encourage a healthy, positive relationship between the

Sonoran Desert and its inhabitants, and the needs of humanity" (*Vista* 1999). Significantly, although they all address conservation and environmental concerns, none of a slate of several mission statements developed for ISDA ever mentions the biosphere reserve concept.

Debate over Incorporating

The issue of maintaining and enhancing participation in ISDA generally revolved around whether the group should remain an open forum or be incorporated under a board of directors. Over the months following the Land Use Forum, the general consensus of the group gradually shifted from the former to the latter. By April 1993, the group agreed by unanimous vote to seek nonprofit status and pursue incorporation for the group. However, some ISDA participants remained concerned over the degree to which a formalized board might inhibit public participation. In June 1993, Propst discussed with the group the specific pros and cons of incorporating (CCR 1993, c.1993a), listing the advantages of incorporating as follows:

- establishing a clear decision-making process with more definitive "official" decisions;
- helping with fundraising (IRS tax exempt recognition);
- establishing a "more permanent organization, with long-term continuity"; and
- allowing the organization to enter into contracts.

On the other hand, he carefully listed the advantages of remaining an informal committee:

- maintaining informal and open participation and decision-making; and
- avoiding corporate reporting requirements and associated costs (Comité Consenso Sonorense Internacional 1993).

Propst emphasized that the election of a board "was part of an evolutionary process and should not be considered a dramatic change. Participation was still open to anyone interested and each individual could bring up issues, had a voice, and a vote" (ISDA 1994c).

After several drafts of the bylaws (CSI 1993a; Quintana Silver 1994b), ISDA was incorporated as an Arizona-based nonprofit organization in January 1994 (Laird and Anderson 1996a), and the first officers of the board were appointed in March 1994 (Quintana Silver 1994a). The idea of formally incorporating ISDA in Mexico was discarded early on, although in 1995 ISDA was registered as a U.S. nonprofit with the full and formal authorization of the Mexican Secretariat of Foreign Relations. That status has allowed ISDA to operate in Mexico—for example, to hold meetings, open bank accounts, hire people—but prohibits it from fundraising activities in the country (García Blanco 1995).

Not long after incorporation, Laird emphasized that despite "the fact that the Alliance is now recognized as a non-profit corporation in Arizona, we have not changed our primary goal: to provide an opportunity for local residents, organizations, and government agency officials to discuss issues, resolve differences of opinion, and develop programs geared at providing a better future for the region" (Laird and Nagel 1994c).

Institutionalizing Diversity on ISDA's Board

In essence, board membership was to be open to "any person who attends an annual meeting of the corporation or a meeting of the Board of Directors" (CSI 1993a). Yet a perennial matter of concern in working out the bylaws was how to ensure adequate representation from all of the highly diverse community members, be they O'odham, Mexicans, Anglos, NGO staff, business representatives, or government officials (Laird 1992). Participants continually pointed to problems in maintaining general interest and participation in the process—particularly in regard to the Tohono O'odham, who had not been traditionally involved in decision making outside their reservation (Laird 1992, 1993a). For example, Laird noted that "at an early meeting of the Alliance held in Sonoyta, there was a long discussion about this region being represented by only two 'nations' (the U.S. and Mexico)" (Laird 1994a). Yet she also attested that

> Cultural diversity, represented by each Alliance member, is ISDA's strength. Each individual brings a new set of eyes and ears, a new perspective and opinion. Each member represents an opportunity for developing innovative solutions and strategies for resolving differences. This idea was so essential to the framework of the Alliance that it, too, was codified in the bylaws. . . . [There is] little hierarchy. It is a highly participatory, open process, geared toward developing active participation from the grassroots level (Laird and Anderson 1996b).

Contemporary accounts also showed that ISDA garnered praise for being well balanced and even "the best thing going on the border right now" (Christensen 1994; SI and FPN, pers. comm. c.1993; Williams 1994). Floyd Flores of the Tohono O'odham Nation (TON) noted that ISDA's "kind of communication is totally new.... We had never talked before" (Williams 1994).

From the first draft, ISDA's bylaws mandated its tricultural character by requiring a minimum of three members in each of the following categories: non-indigenous resident citizens of Mexico, non-indigenous resident citizens of the United States, indigenous resident citizens of the United States, and indigenous resident citizens of Mexico. Additional board members, it was added, "need only be certified ISDA members" (ISDA 1996; Laird and Nagel 1994b). This mandated mix of cultures on the board became one of the "signatures" of ISDA, mentioned in nearly all media coverage of the initiative.

While the bylaws stated that board members did not officially represent any institution in their capacity as board members, there was still concern over the proper role in ISDA for government officials. In 1993, ISDA participants discussed the idea of a government advisory committee, which would include federal, state, county, and tribal governmental officials serving "in an advisory capacity as U.S. law generally prohibits their participation in non-governmental organizations as voting members" (Laird 1993i). Such an advisory committee was established in the bylaws of the Alliance, which state that the committee would have a voice in ISDA's proceedings and activities, but have no vote in ISDA's board (Laird 1994e). The committee was to include representatives from ORPI, the U.S. NPS Mexico Affairs Office, Cabeza Prieta NWR, BLM, Luke Air Force Base, Arizona Game and Fish Department, Centro Ecológico de Sonora, U.S. Immigration and Naturalization Service, U.S. Border Patrol, U.S. Marine Corps, and Pima County Planning Division.

The handful of government officials involved in ISDA at the time generally professed a very positive attitude toward it. While this was partly due to the serendipitous congregation of open-minded personalities in those government positions, it was probably more a result of the fact that, as Laird put it, "every blessed one" of the agencies was conducting or was about to conduct "planning" exercises for their respective land. ISDA offered a way for those land managers to reach out to the local communities —particularly for those officials who lived outside the region (e.g., in Phoenix) or within their "glass jar" protected area. And as former Cabeza Prieta NWR manager Bob Schumacher told me, ISDA was attractive to his fellow government officials because "we can do things with an NGO we can't do in a competitive arena between bureaucrats.... An NGO can bring the bureaucrats together, sometimes where even a strong personality bureaucrat can't because we are constrained ... and held back by our respective agencies." Furthermore, he said, both he and Harold Smith, ORPI's superintendent, "were personally very much loud proponents" of a biosphere reserve approach to land management, and ISDA appeared to be in a position "to make the agencies come up with an umbrella master plan—which is what a biosphere essentially would be."

The SI-ORPI Cooperative Agreement and ISDA's Early Programmatic Agenda

Even as ISDA was coalescing into a formal entity, its leadership was finding it difficult to coordinate participation and move the agenda forward. In a March 1993 memo to participants, Laird (1993e) recorded her frustrations with "so many pertinent issues, so much to do, and not enough time," and said that "the process was going so slow." Lightly reprimanding herself ("Nagel suggested that I should quit being so 'Anglo,' that these things take time"), she noted that "we are light years ahead of where we were just 6 months ago" and pointed to the success of the 1992 Land Use Forum, high participation in steering committee meetings, a bibliography on the Sonoran Desert, the commencement of a resource inventory, and planning for another major conference in 1994. She also noted that the group's mailing list contained over one hundred people interested or working in the area. "Participation continues to grow!" she reported. "We are receiving attention."

Laird had good reason to feel optimistic about the attention ISDA was receiving not only within the region, but also from outside—including from powerful decision makers. The same month she voiced her frustrations, Laird, Propst, and a small group of ISDA participants traveled to Washington, D.C., for a series of meetings with staff members from the offices of Senator Dennis DeConcini (D-AZ), Representative Jim Kolbe (R-AZ), and Representative Ed Pastor (D-AZ). Laird noted the enthusiastic reception to the idea of an international biosphere reserve in the Sonoran Desert, pointing out that the staff from the offices of Senator DeConcini and Representative Pastor were "particularly keen" on the ideas of creating an international biosphere reserve and of "enhancing international relationships between the Tohono O'odham, U.S., and Mexico" (Laird 1993f). The meeting concluded after DeConcini's staff requested a letter outlining ISDA's approach and budget needs (Laird 1993d, 1993f, 1993j).

Immediately after those meetings, ISDA participants drafted a proposal to Senator DeConcini that came up for review and approval at an April 1993 meeting. According to Laird's notes, the purpose of the proposal was "to pursue a feasibility study to determine the possibility of creating a binational park, protected area, biosphere reserve, whatever the name" (Laird 1993f). The proposal led to a congressional appropriation of $300,000 to ORPI in 1994, $250,000 of which was to be used for ISDA-related projects (Condo and Bishop 1995). However, at least partly because ISDA had yet to build up an internal structure for financial accounting, the finances were channeled through SI via a Cooperative Agreement with ORPI signed in August 1994 (Organ Pipe Cactus National Monument and Sonoran Institute 1994).

Notably, although funding for the Cooperative Agreement had its roots in discussions over an international biosphere reserve for the region, the Cooperative Agreement itself did not center around biosphere reserve concepts—indeed, the word *biosphere* is not mentioned a single time in the text of the agreement. Rather, according to Laird and Nagel (1994a), the Cooperative Agreement reflected "the priorities set by the Board at their first meeting and provides a framework for our work." In other words, the structure of the Cooperative Agreement indicated the extent to which Laird and Nagel worked to keep ISDA open to *all* issues of concern to ISDA participants—and not just to issues that would traditionally pertain to biosphere reserves. Their efforts paid off, for during that time period the minutes recorded lengthy discussions over health care issues, border crossings, NAFTA issues, educational curriculum development, highway construction, and other issues that simply do not fall under the rubric of traditional biosphere reserve thinking (CCI 1993c; CCR 1993; ISDA 1993b, 1994c; Laird 1992, 1993a, 1993b, 1993h, 1993i, 1993j; Nagel 1994; Quintana Silver 1994b). Naturally, many of the priorities set by the board were incorporated into ISDA's early programmatic agenda, the major components of which are listed under the following headings.

Conferences and Meetings
By the spring of 1993, Laird and Nagel had identified public community forums as a central task of the "Comité Consenso." A main objective of these forums was to assess "community assets and values" (SI and FPN

c.1993). In 1994, eight ISDA general meetings were held in the communities of Caborca, Sonoyta, Puerto Peñasco, San Luis, Ajo, and Gila Bend, with participation averaging forty people per meeting (ISDA 1994a). Internal ISDA documents stated,

> Perhaps our most important accomplishment to date has been to establish an on-going dialogue among border community residents and government agency representatives. ISDA is seen as a "safe place" where diverse people can share their experiences and knowledge, grapple with local concerns, and resolve critical issues. It is also a place where goals can be set, objectives met, and positive results established. (ISDA 1994a)

Nearly everyone I spoke with who was involved in ISDA during this period believed that the community meetings successfully opened lines of communication between previously isolated and self-contained segments of the Sonoran Desert community.

1994 Bridging Borders Conference

Planning for a follow-up to the 1992 Land Use Forum eventually led to a second major international conference that took place in January 1994 in Puerto Peñasco, Sonora. The principal objective of the conference was to develop a "vision statement and plan of action" for meeting the "economic needs of communities in the western Sonoran Desert U.S.-Mexico border region while guaranteeing respect for cultural heritage and traditions, and ensuring protection of natural resources" (Keiller 1993; Laird 1993e, 1993f, 1993h). As with the 1992 Land Use Forum, the objectives of the 1994 Bridging Borders conference were expansive, covering the three broad themes of economic development, social and cultural issues, and environmental issues (Laird 1993i).

Sponsored by over thirty-six organizations and attracting over three hundred individuals from the U.S., Mexico, and the Tohono O'odham Nation (ISDA 1994b), the conference "marked the formal establishment" of ISDA through the first board election (ISDA 1994b; Laird and Nagel 1993a, 1993b). U.S. Senator Dennis DeConcini, one of the several dignitaries at the conference, described ISDA as a "model for cross-border cooperation and international problem solving. It represents the kind of long-range, community-based planning necessary for protecting shared ecosystems in the border region" (Laird 1994c).

The heart of the conference consisted of roundtable workgroups and panels, from which recommendations were incorporated into an Action Plan that set out numerous recommendations under the seven broad categories of health care, trade and transportation, environmental education, planning for protected areas, information exchange, tourism, and economic development. Despite its audacious if not preposterous enormity, the Action Plan both reveals the breadth of ambition ISDA had adopted at this point and appears to have been the first comprehensive attempt to prioritize and classify the principal environmental and economic challenges facing the region (ISDA 1994b).

In addition to addressing those challenges, the conference was a cultural exchange, with presentations by indigenous cultures of the region (ISDA 1994b). A session titled "Stories of the Region" featured Gary Nabhan, Tohono O'odham traditionalist Danny Lopez, Sonoyta historian Jorge Luis Gamboa, Puerto Peñasco historian Guillermo Munro, and Bill Hartmann, the author of a major treatise on the natural and cultural history of the Sonoran Desert (Hartmann 1989; ISDA c.1994). Scientists Richard Felger, Alberto Búrquez, Carlos Valdez, and Carlos Castillo led a "Science & Research" session featuring presentations on "binational research activities aimed at furthering our understanding of this complex and fragile desert environment" (ISDA 1994b, c.1994). Conservation activist David Foreman of the Wildlands Project (noted earlier in regard to Earth First! and the mythology of the Pinacate) "unfolded his dream-map of wildlife corridors that might connect the remaining wildlands" (Steffens 1994). The session emphasized the "need for detailed ecological monitoring," particularly for Sonoran pronghorn (*Antilocapra americana sonoriensis*) and ironwood (*Olneya tesota*) (ISDA 1994b).

As one reporter noted, the conference may not have seemed "like such a big deal, but following so many years of autocratic decision-making by governments that treat the border as a problem to be controlled and separated, many area residents weren't even sure how to begin a dialogue, much less one conducted in three languages" (Steffens 1994). While Laird viewed the 1994 Bridging Borders conference as a "tremendous success" for ISDA (Laird and Nagel 1994d), one of its most significant results was the establishment of the Sonoran Desert Taskforce, a group of activists and scientists distinct from ISDA that, as described extensively below, would come to play a critical role in promoting a biosphere reserve designation for the western Sonoran Desert.

Regional Resource Inventory and Regional Profile

In 1992, ISDA participants discussed the idea of a regional resource inventory for the western Sonoran Desert border region. Laird and Nagel saw it as a way of providing a much needed foundation of baseline information "for future decision-making and the development of a strategic plan of action that transcends the international border" (SI and FPN c.1993). The primary purposes of the resource inventory were to identify and categorize natural, cultural, and economic regional resources of the region and to aid in the development and understanding of a regional approach to land use and conservation. By January 1993, a rough draft was available (Laird 1993b), and in March 1993, Laird noted that "University of Arizona and Mexican researchers have begun compiling information for a resource inventory which will provide a basis for defining what we have and where we want to go. . . . No other effort like this exists—which compiles economic, and natural and cultural resources into a single volume—on the border" (Laird 1993e).

Apparently, the science-based resource inventory evolved into the idea of a more public "regional profile" document. In November 1993, a draft Regional Profile was being circulated for review (Laird and Nagel 1993a), and the project was still underway in June 1994 when it was described as "an easily accessible, well-illustrated succinct report identifying the region's natural, cultural, and economic resources with regional maps utilizing Global Information System (GIS)" (Laird 1994c). Unfortunately, I was unable to find a copy of any version of the report and have found little indication that the report was ever finalized, published, or used for any substantive informational purpose (although see ISDA 1995a).

Juntos

By the spring of 1993, ISDA had identified Juntos: Maestros y Niños del Desierto as a priority project (SI and FPN c.1993). Described as "a unique model for binational cooperation in environment education," (SI and FPN c.1993) the Juntos program was the first bilingual and multicultural environmental curriculum to foster "awareness of natural and cultural history of the Sonoran Desert" as well as "a sense of place and land stewardship, especially in regards to protected areas" (Pearson 1998, 21–22). With a hands-on approach to history, math, science, and art for local elementary students, the curriculum taught students about "the unique cultural heritage of the region, human impact on the natural environment, and land use changes over time" and encouraged "greater community participation in a tri-cultural setting" (ISDA c.1996a, c.1996b; SI and FPN c.1993).

In 1994, a final draft of the curriculum was compiled and translated into Spanish (ISDA 1994a), and the program has since been implemented in a number of communities. Supported early on by the National Park Foundation through a grant to ORPI, the program was permanently transferred to ISDA in 1995, with ORPI providing initial staff and interpretation (Pearson 1998, 22; SI and FPN c.1993). However, after some wrangling over ownership of the program, it was ultimately transferred to the Environmental Education Exchange (Environmental Education Exchange n.d.).

Roots-Raices-Ta:tk

Laird (1994f, 1994d) described the Roots-Raices-Ta:tk Youth Group program as "a community-driven native seeds planting project" and "a cross-border exchange program geared at high school students and community residents." Initiated in 1994, general activities of the program included developing leadership skills, promoting environmental awareness, and communicating the importance of cultural diversity of the Sonoran Desert's inhabitants. Some specific activities of the program included the monitoring of "light pollution" at the Kitt Peak National Observatory on the Tohono O'odham Reservation, environmental restoration activities in the Ajo area, a resource inventory and data collection management project at Cabeza Prieta NWR, studying ecological restoration in El Pinacate Biosphere Reserve, and recycling projects (ISDA n.d.).

Other Activities

In addition to the preceding activities, ISDA participants were involved in an array of other projects. The broadness—if not the depth—of those projects serves as a good indication of the expansive territory ISDA was trying to cover during its early years:

- *Health issues.* Through extensive advocacy and consultation with government agencies, ISDA addressed—and, in a few cases, solved—problems relating to border crossings and health care.
- *Economic analysis.* In 1994, ISDA completed the first draft of an economic analysis to "assess the historical, present, and projected economic activity in this tri-cultural, binational

region, including the role of tourism and other 'clean' or non-extractive industries, and outline alternatives for sustainable economic development" (ISDA 1994a, 1995a; Laird 1994c). That draft may have been the precursor of a series of small community meetings "to identify local environmental and economic development priorities" that was organized in 1995 by ISDA and SI (Laird et al. 1997, 310). The outcome of those meetings was a regional working group to assess the feasibility of establishing ecotourism ventures in the Sonoran Desert. The group's work in turn constituted a major impetus to the establishment of the tourism venture *La Ruta de Sonora*, which gives travelers an in-depth perspective of the Sonoran Desert cultures and environment (see www.laruta.org).

- *Specific campaigns.* Though generally shying away from direct advocacy on overtly political issues, ISDA has at times given its imprimatur to various specific campaigns, such as one to protect the instream flow of the Colorado River (ISDA 2000a, 2000b) and another to promote "smart growth" in Arizona (ISDA 2000c).

- *Mining.* Many ISDA participants were involved in protests against a proposal by the mining company Minera Hecla to open a mine in Quitovac, Sonora. Several ISDA meetings covered the issue in depth, at least one with participation from Minera Hecla officials. The debate ended up embroiling ISDA in the contentious relationship between the company and its detractors. Ultimately, the mine was opened, but by December 1998, reserves at the mine were exhausted, and mining ceased (Hecla Mining Company n.d.; Ortiz Garay 1999; Propst 1994).

- *Highway expansion.* ISDA has served as a forum for discussion over the management and potential expansion of Highway 85, the sole north-south auto route running through the center of ISDA's area of concern. In a 1994 listing of its accomplishments, ISDA reported that it had "successfully campaigned" to include Highway 85 in the "twenty-year plan" of Arizona's Department of Transportation, which, it was hoped, would eventually "decrease road hazards and include mitigation measures for wildlife" (ISDA 1994a).

- *Educational Infrastructure*. In 1994, ISDA held five meetings of the Border Environmental Education Group, assisted in coordinating and convening a public hearing with federal officials to discuss the NAFTA-related Border Environmental Cooperation Commission (BECC), and participated in a "mapping exercise" sponsored by the Wildlands Project (ISDA 1994a).
- *Regional interpretive center*. ISDA's board spent a significant amount of time investigating the possibility of establishing a regional interpretive and research center (CCR 1993). Several case studies of nature centers in other regions were presented and discussed at ISDA meetings, including a close look at the National Big Horn Sheep Center in the community of Dubois, Wyoming (CCR 1993; Laird 1993e). Although ISDA went so far as to plan a feasibility study (ISDA 1995a), the idea was eventually downgraded to a Regional Resource Center, which was housed within ISDA's administrative office in Ajo, Arizona. The center remains today a bilingual clearinghouse and lending library for government, popular, and scientific information about the region (ISDA c.1996a, c.1996b).

Designation of the Pinacate Biosphere Reserve

Twelve kilometers (seven and a half miles) to the northwest of Puerto Peñasco in the *Sierra del Pinacate* lies the volcanic Cerro Prieto—or Black Hill—overlooking the Gulf of California. There, on June 10, 1993, Mexico's President Carlos Salinas de Gortari announced, "We have decided that here ecology has priority over politics" (La Rue 1993). The comment came during a declaration ceremony for the Pinacate and Gran Desierto de Altar Biosphere Reserve and the Upper Gulf of California and Colorado River Delta Biosphere Reserve. The occasion brought together numerous other dignitaries: U.S. Secretary of the Interior Bruce Babbitt, Mexican Minister of Social Development Luis Donaldo Colosio, Arizona Governor Fife Symington, the president of the University of Mexico, and the director of the Mexican Institute of Ecology. Also present was Sylvester Listo, the chairman of the Tohono O'odham Nation, and a delegation of about three dozen O'odham (Laird 1993i).

Various ISDA participants and observers have, in casual conversation, credited ISDA for providing the impetus for the Mexican designation of the Pinacate as a biosphere reserve. At least two pieces of evidence, however, make that assertion suspect. First, there is no substantive documentary evidence (including notes, letters, and meeting minutes) that ISDA had any official or unofficial input into the decision to designate the Pinacate as a biosphere reserve. Second, the historical record points to other actors and influences that were instrumental in obtaining the designation. Laird, for example, emphasized that The Nature Conservancy had been working on protecting the Pinacate for years and that the designation "would not have happened without them." Castillo-Sánchez (n.d.) argued that studies conducted on the threats to the endangered Sonoran pronghorn were "key factors" in the designation of the Pinacate biosphere reserve. Steinhart and Blake (1994) and Búrquez (1998, 75) credited work of the Centro Ecológico de Sonora and the Universidad Nacional Autónoma de México (Estación Regional Noroeste, Sonora) for the biosphere reserve designation. Another influential institutional actor was the Centro Intercultural de Estudios de Desiertos y Océanos (CEDO) of Puerto Peñasco, although its efforts primarily focused on the establishment of the Upper Gulf Biosphere Reserve. But given that these designations were made at the federal level under the notoriously secretive modus operandi of the Institutional Revolutionary Party (PRI) (Oppenheimer 1996), undoubtedly a great deal more happened behind closed doors than will ever be found in the historical record.

Yet despite the lack of evidence of ISDA having played any direct role in the designation of the Pinacate Biosphere Reserve, it nonetheless appears to be the case that the *effort to establish ISDA* was a key factor in allowing the designation to occur. The distinction is subtle but important and bears repeating: Much of the impetus toward the designation came not from ISDA per se but from personal and political connections made during the formation and development of ISDA. The connection between the creation of ISDA and the designation of the Pinacate was largely revealed to me by biologist and government official Ezequiel Ezcurra, one of the principal scientists who had investigated the Pinacate for possible designation as a biosphere reserve during the early 1980s.

In early 1992, Ezcurra was advising graduate students at the Universidad Nacional Autónoma de México. One day the Mexican Minister of Social Development Luis Donaldo Colosio invited him to the Ministry's offices. "He invited me for coffee," recalled Ezcurra, "and, you know, you don't say no to a minister when he calls you down and says, 'Would you like to come and have coffee?' I mean, you just go." At that time, Colosio was running for president as the principal candidate under the ruling PRI. Although Ezcurra had some reservations about Colosio and the PRI—particularly regarding their indigenous policies—he found Colosio to be "very honest, very sensitive towards environmental problems, and wanting to do things." Colosio asked Ezcurra to be his general director of natural resources, waving off Ezcurra's concern over being Argentinian-born. Thus, Ezcurra, a naturalized Mexican, became one of Mexico's first foreigners appointed to a high-ranking government position.

One of Ezcurra's first undertakings in office was to attend the 1992 Land Use Forum in Ajo. At the bed and breakfast where he was staying, he informally met with some of the O'odham and Sonoran government officials (including Maria Elena Barajas, the Sonoran government minister of the environment, and Dr. Samuel Ocaña, the former governor of Sonora and at that time the director of the Centro Ecológico in Sonora), as well as Susan Anderson of The Nature Conservancy. Ezcurra said to the group, "'How about we move forward to protect el Pinacate? . . . You know, Colosio's from Magdalena, so he would support us. We have all been dreaming of a protection of the Pinacate for different reasons—all the way from sacred sites to the protection of the natural environment. NAFTA is starting to make progress, and putting a protected area along the border has ceased to be a taboo.' And they all said yes, let's do it."

Ezcurra returned to Mexico City, whereupon Colosio informed him that he was looking for "something effective that we can do quickly, but will draw attention on our interests in preserving the environment." Ezcurra promptly suggested that Colosio protect the Pinacate and the Upper Gulf of California as biosphere reserves:

> And he said, "Why?" I said, "Well, it's a border protected area. And those are taboo in Mexican administrations, so we'll really make a splash. Secondly, everybody wants them. It's going to make you popular. There is practically nobody that really opposed the creation of a protected area there. It's an easy one." And I said, "Third, it can give Salinas"—who was a very special character—"a lot of visibility vis-á-vis NAFTA and the free trade agreement negotiations."

Ezcurra said that Colosio "loved the idea because he was from Magdalena, he was from Sonora. He knew these places." Still, Colosio expressed concern over what kind of support he would garner for such an action. Because "Colosio gave a tremendous importance to social groups," he specifically asked if there was "anybody backing" the idea. Ezcurra responded that both the O'odham and ISDA were backing it. "It won't be contested," Ezcurra said. "It won't be rejected by social groups—there is widespread support." Colossio responded positively: "Okay, do it, but keep that support going."

Ezcurra was also influenced by the arrival in Mexico City of a group of U.S. and Mexican Hia-Ced O'odham, many or all of whom had also attended the 1992 Land Use Forum. They had driven all the way down from the border, recalled a still impressed Ezcurra, in a crowded "derelict" van simply in order "to have their voice be heard." He hosted the Hia-Ced, taking them around Mexico City, particularly to the shrine of the Virgin of Guadalupe. (Ezcurra noted that while the O'odham are not strictly Catholics, a "religious syncretism" in O'odham culture perceives the Virgin of Guadalupe "as a central force in their native identity or their identity as native peoples.") While the Hia-Ced seemed to sincerely appreciate this personal attention, they were more impressed by Ezcurra's incorporation of the O'odham into the planning group for the Pinacate protected area—and by his committing the government "to give them full and ample participation in the planning of this protected area." After a formal agreement was signed by both Ezcurra and the O'odham, the latter traveled back to Mexico City several times to participate in planning for the Pinacate.

Not having the time to work on a proposal for the Pinacate himself, Ezcurra went back to his colleagues Búrquez and Castillo, asking them to update the proposal they had collaborated on in the early 1980s. He pointed out that all their previous scientific research could now be incorporated into a new suite of technologies—particularly geographic information systems (GIS)—that had been developed since the 1981 proposal. Búrquez and Castillo agreed to work on it, and the proposal was completed in April 1993 (Búrquez and Castillo 1993).

The proposal was then presented to President Salinas, who at the time was heavily invested in pursuing the success of NAFTA (and was not yet mired in the scandal that would eventually ostracize him from the country). Salinas's interest in creating new protected areas near the U.S. border could well have been related to his desire for the United States to sign the NAFTA agreement (La Rue 1993). Like Colosio, he was interested in whether the proposal had popular support from *grupos sociales*, and Ezcurra was able to assure him that it did:

> I know "popular" sounds like a pompous word, but there was generalized support for the project. I mean, the fact that the Tohono O'odham said, yes, this is a good project, we want it; the fact that Centro Ecológico said, yes, this is great; the fact that The Nature Conservancy in Tucson said yes; and the fact that ISDA pressed to get this going—all these things started something up. And they got enough momentum that if you just left an open door, you took down the barriers that existed inside the federal government in the past.

Ezcurra emphasized the latter point—that his role was simply to open the door to Colosio's office—"to unlock the door from inside for them to do it."

Once the Mexican government had designated the Pinacate Biosphere Reserve, its next step was to generate a management plan for the buffer zone areas (Laird 1993j). O'odham participation in development of the plan—as called for in their agreement with Ezcurra—apparently began well, and O'odham representatives regularly reported to the ISDA board on their role in creating management plans for the newly designated Pinacate Biosphere Reserve. For example, in April 1994, two O'odham participants in ISDA reported to the board that they had submitted a set of principles regarding the management of the Pinacate to the government of Mexico," and that Joe Joaquin of the Tohono O'odham Nation (and future president of ISDA's board) had been appointed O'odham representative for the Pinacate initiative (Quintana Silver 1994b).

Peggy Turk Boyer, executive director of the Centro Intercultural de Estudios de Desiertos y Océanos in Puerto Peñasco, told me that "the O'odham culture was very much involved in the meetings" on developing the Pinacate's management plan, and that the plan incorporated "the O'odham view of the Pinacate." But even though their participation resulted in "a lot of respect developed for the O'odham group," the collaboration with the Mexican government ultimately broke down for a number of reasons. First, as both Ezcurra and Turk Boyer pointed out, the two "partners" were politically and culturally worlds apart. Furthermore, although the Pinacate Biosphere Reserve had even hired an O'odham at one point, the Pinacate managers simply could not find either Mexican or U.S. O'odham that would be "really available" to live and work in the Pinacate. As Lorraine Eiler pointed out to me, in the late 1800s the O'odham living in the Pinacate had been "rounded up" and "dumped" in the town of Caborca on the southeast edge of ISDA's geographic coverage (see Figure 3.1). Consequently, as Ezcurra put it, most O'odham "have only heard about the Pinacate from their elders, and they cannot recognize the sacred sites once you go with them afield."

Ezcurra described "two interpretations" of the O'odham's contemporary ignorance of their sacred sites. The "reactionary interpretation," widespread within the Mexican government, held that "these guys don't even know their sacred sites; we shouldn't even take them into consideration in the planning of a protected area." In contrast, Ezcurra wanted to instill the "culturally sensitive" interpretation that the O'odham had "lost contact with the sacred sites because of forces that are totally external to them"—and that the Mexican government ought to reconnect the two. But Ezcurra's chance to bring that second interpretation to fruition was brought to an abrupt halt by violent politics at the national level in Mexico. For on March 23, 1994, Ezcurra's boss and the leading presidential candidate, Luis Donaldo Colosio, was assassinated in Tijuana.

Pointing out that Colosio had been "totally a part of the PRI establishment," Ezcurra self-consciously avoided "romanticizing him excessively." Yet Ezcurra did feel strongly that Colosio "was different" and sincerely believed in changing the system from the inside. "When Colosio died, I just lost my compass," said Ezcurra:

> The whole dream just—the whole bubble just went to pieces. So perhaps this is a story with a sad ending, because a lot of the dreams that started that day when we declared the Pinacate—I thought that it was just the beginning of a new era, in which . . . proper attention to the issues of Indian affairs and proper attention to the issues of environmental degradation were coming into decision making of the federal government in Mexico.

Ezcurra concluded his story by reservedly pointing out that after he left office, the Mexican government's policies toward indigenous peoples changed for the worse. "You know," he said, tongue somewhat in cheek, "every governmental officer tends to think that their time is a glorious one and the ones that came after him or after her were really lousy. But I understand that the current director of protected areas doesn't have the sort of pro-indigenous people sensitivity that I think I had.... I get the impression that the present authorities in Mexico don't have the same interest in the participation of native peoples and conservation."

Looking back at this winding history, it is fair to say that Ezcurra's presence at the 1992 Land Use Forum set the stage for his subsequent political maneuvering to have the Pinacate officially designated as a biosphere reserve. In other words, had there been no effort to create what would eventually become ISDA, it is doubtful that the chain of events leading to the Pinacate Biosphere Reserve would have occurred. ISDA participants clearly recognized Ezcurra's key role, and a few months after the dedication ceremony for the Pinacate Biosphere Reserve, they presented him with the Julian Hayden Award at its 1994 Bridging Borders conference.

In sum, Mexican federal politics (including a tragic assassination) had largely stymied the hopes of the O'odham and ISDA for meaningful indigenous participation in management of the Pinacate Biosphere Reserve. As will be described in the following sections, negative forces at the federal level—this time in the United States—would strike again at the efforts of ISDA.

Expanding Biosphere Reserves in the Western Sonoran Desert

Between the summers of 1993 and 1994, ISDA was developing rapidly through a series of well-attended forums and a number of burgeoning programs. But even as the initiative gathered steam under its own momentum, Laird and Nagel were still very much guiding the process. Theirs was something of a high-wire act, not because they were in the limelight—something both avoided—but because they found themselves balancing between two countervailing forces. On one side were those pushing hard for an international biosphere reserve designation for the western Sonoran Desert; on the other side, some board members remained skeptical or unsure of what "biosphere reserves" were all about.

As described previously, Laird and Nagel eschewed the idea that anybody—including themselves—could march into the region to announce that local residents had to "save the desert." They would have likely agreed with what Hia-Ced O'odham Lorraine Eiler told me, that too great a focus on biosphere reserves would have been the wrong approach for ISDA:

ISDA was just starting to have town hall meetings in most of the communities within the Sonoran Desert, such as Gila Bend, San Luis, Puerto Peñasco, Caborca and the Tohono O'odham Nation. . . . And the Man and the Biosphere was the last thing they wanted to talk about because we were still in the "my home is this, my home needs this, or I need to feed my family." We were just starting out, and we felt that if the push was on for the Man and the Biosphere, to go forward would just ruin everything, all our work; it would just completely do away with the concept of what ISDA stood for before it was ISDA.

Laird also pointed out to me that "the Ajo community was still seething from being restricted from using Cabeza when it became wilderness" under the 1990 Arizona Desert Wilderness Act, a law that had been championed by several national conservation groups. Coming into Ajo two years later, Laird and Nagel nonetheless had to grapple with the community's perception of them as "outside" environmentalists.

In essence, Laird and Nagel believed that given sufficient time, ISDA would ultimately engender the most democratic and solid foundation upon which to implement biosphere reserve concepts in the Sonoran Desert. Consequently, they consciously avoided pushing biosphere reserve concepts on ISDA participants and were careful—at times, painstakingly careful—not to subvert the board's decision-making authority. Accordingly, a subtext running through the minutes of the first two years of ISDA's meetings is a deliberate effort not to dominate the proceedings with discussion of biosphere reserves. Notably, in thirty-eight pages of minutes from twelve meetings between 1992 and 1994, the phrase "biosphere reserve" comes up approximately ten times, typically in reference to what had been going on in regard to biosphere reserves both in Mexico and the United States. So although biosphere reserves were discussed, meetings were decidedly not *about* them. Furthermore, when the group generated lists of priority issues during that time period, biosphere reserves were typically located at the bottom (CCI 1993c; Laird 1993i; Portman and Laird n.d.; Sierra Club Grand Canyon Chapter 1993).

Yet even as Laird and Nagel strove to keep ISDA open to a broad swath of societal issues and concerns, both had been in continual communication with national and international MAB officials since before the 1992 Land Use Forum, and both relied on biosphere reserve concepts when presenting ISDA to outside audiences. Increasingly, they were not alone since many other ISDA participants picked up the banner for biosphere reserves during this period. Already by the end of 1993, when Laird, Nagel, Tony Ramon (a Tohono O'odham representative on ISDA's board), and ORPI superintendent Harold Smith represented ISDA at a Strategic Planning Conference of the U.S. MAB Program that was held just outside of Rocky Mountain National Park and Biosphere Reserve, it had become clear that ISDA was inextricably entwined with the MAB Program. Presented at the conference "as a model and case study in community participation," ISDA was subsequently featured as one of twelve biosphere reserve case studies in the U.S. MAB Program document *Biosphere Reserves in Action: Case Studies of the American Experience* (Condo and Bishop 1995).

Laird and Nagel's balancing act was complicated by outside pressure to engage ISDA as an advocate for an international biosphere reserve designation. The pressure came from a number of like-minded conservationists and scientists working in the region, the most prominent of whom was author and biologist Gary Paul Nabhan. Recipient of the 1986 John Burroughs Medal for outstanding natural history writing and, in 1990, of both a Pew Scholarship for Conservation and the Environment and a MacArthur Fellowship (the so-called "genius grant"), Nabhan has been described as "one of the leading figures in the field of environmental literature" (Slovic 2004, 78, 90–91). As early as the 1992 Land Use Forum, Nabhan had been collaborating with other individuals to push for an international biosphere reserve that would encompass most of the western Sonoran Desert. ISDA's Bridging Borders conference in January 1994 apparently inspired him to convene "a loose group" of scientists, O'odham representatives, NGO staff, and "desert rats" in Tucson the following month. Adopting the title of Sonoran Desert Taskforce, the group undertook to draft a proposal for "cooperative management and resource protection" of the western Sonoran Desert (Laird et al. 1995). Holding its periodic meetings at Nabhan's institutional home of the Arizona–Sonora Desert Museum in Tucson, the Taskforce focused on designating an "international biosphere cooperative program" that would designate "some portion of the area as an international biosphere reserve," as well as "formalize the type of public-private partnership pioneered by the International Sonoran Desert Alliance" (Nabhan et al. n.d.; TPSDBR 1994). The Taskforce articulated the distinction between itself and ISDA in describing itself as "a temporary, task-oriented coalition of scholars, activists and non-government organizations, whereas ISDA [was] the only grassroots and community-based group in the Pinacate/OrganPipe [sic] vicinity" (TPECR 1994).

Although the Taskforce would come to have a significant impact on ISDA, the route its influence took was not straightforward. On the one hand, Laird recalled, the Taskforce was "impatient and unwilling to wait long for the relationship [with ISDA] to build," and both she and Nagel continued working to keep ISDA as open a process as possible. On the other hand, Laird, Nagel, Propst, and Superintendent Smith attended at least one of the Taskforce's meetings, and they all fundamentally agreed with the Taskforce's basic premise that an expanded biosphere reserve would be a positive development in the region. Thus, in their correspondence with the board around the summer of 1994, Laird and Nagel began to emphasize the potential for biosphere reserve concepts in the western Sonoran Desert. This was also when the Taskforce—represented mostly by Nabhan—began to meet directly with the ISDA board.

At this point, the ISDA board as a whole had become engaged in the application of biosphere reserve concepts to the western Sonoran Desert. Yet it is crucial to understand that there were two somewhat distinct efforts going on simultaneously. On the one hand, there was the proposal to complete an international Sonoran Desert biosphere reserve—the main objective of the Taskforce. On the other hand was an attempt to establish a collaborative agreement between the U.S. agencies under which ISDA would act as a MAB "partner" or "cooperative program." Although the two were closely entwined, it is worth looking at each of them in turn— beginning with the latter, as this was where Laird and Nagel were focusing their energies.

A Biosphere Cooperative for the U.S. Side

ISDA's files contain a draft international agreement marked July 1994 between the U.S. government agencies, ISDA, and Friends of ProNatura to establish a "Man and the Biosphere Cooperative" (Anonymous 1994b). Riddled with inconsistencies—most notably the failure to list Mexican agencies as parties to the proposed *international* agreement—the draft apparently went nowhere. However, the draft does reveal just what was being envisioned through eleven formally worded perambulatory clauses ("Considering that . . . "), two of which describe the nature and functions of ISDA and SI, while another refers to the ORPI-SI Cooperative Agreement. Substantively, the draft envisions the following:

- an executive committee with a trinational makeup similar to that of ISDA;
- a Governmental Coordinating Council made up of representatives from each governmental agency;
- a coordinating office; and
- collaboration between federal, state, and municipal entities on natural resource and ecosystem management, culturally sustainable development, research initiatives, information dissemination, and culturally sensitive educational programs

Who wrote this agreement—and why there was apparently no follow-through—is unknown, but it is warranted to speculate that either the drafters realized that the agreement was not yet politically feasible, or they saw it as too large a step for ISDA at that time. Whatever the case, any effort to formalize the agreement seems to have been dropped.

In August 1994, Laird and Nagel wrote to the board to suggest that the "next major step" for ISDA was the "development of a biosphere cooperative program *on the U.S. side* of the border" (Laird and Nagel 1994a). Noting the tangible federal support for ISDA's activities and emphasizing that the "beauty of the biosphere reserve program is its flexibility," Laird and Nagel wrote,

> The biosphere cooperative program would institutionalize the Alliance's current structure (i.e., community representatives from the U.S., Mexico, and Tohono O'odham Nation working as equal partners with a government advisory council). *It would not be a separate group.* The directors of the cooperative would set policy and jointly guide activities for the biosphere reserve program in the U.S., including economic development activities, environmental education, tourism development, joint research efforts, etc. Similar to what is being done now (Laird and Nagel 1994a).

Such a biosphere cooperative program, they continued, would provide "recognition that the Alliance is a force in the region," institutionalize "the current framework" of ISDA, "allow for access and better coordination with the biosphere reserve in Mexico," establish an "internationally recognized structure" that would not change current federal or private lands management, and create the potential for federal and foundation funding. Ultimately, Laird and Nagel argued that the biosphere concept could succeed only through a nongovernmental structure such as ISDA's, and that "if the cooperative is not supported locally and if it is seen by residents as a federal *government* program, then it will not work. This concept must be generated by people in the region. In fact, if we do pursue the biosphere cooperative idea, then it will be you who must sell it to your local communities. So, your perceptions are key." They were careful to note at the end of the memo: "Whatever the Board wants to do regarding this idea is fine by [us]. To do nothing is also an option" (Laird and Nagel 1994a).

The board apparently wanted to do something. One month later, in September 1994, a letter from ISDA's board addressed to "Community Members of the Sonoran Desert" (meaning, most likely, ISDA's membership and participation list) invited recipients to participate in strategizing over several ongoing regional planning initiatives—most significantly in terms of pursuing "a single effort that examines the region as an integrated whole" (ISDA Board of Directors 1994). The letter noted that an "international cooperative biosphere reserve program" is one such alternative, and attached to the letter was a two-page question-and-answer sheet on biosphere reserves in the western Sonoran Desert.

By 1995, the "partnership" had become sufficiently recognized to be included in *Biosphere Reserves in Action*, the document resulting from the aforementioned 1993 MAB conference at Rocky Mountain National Park and Biosphere Reserve. Drafted by a MAB consultant but based on information from Nagel, Laird, Propst, and ORPI Superintendent Harold Smith, the document describes ISDA as a mechanism for expanding the biosphere program to involve an increasing number of regional interests:

> ISDA serves as a regional cooperative that empowers its members to achieve community goals compatible with maintaining a healthy desert ecosystem.... [It] offers a forum for building a common understanding of the ecosystem and for resolving conflicts associated with the management and development of resources. By encouraging communication among people who have traditionally not spoken with each other, ISDA is building recognition of the biosphere reserve as a meaningful concept of practical value in their lives. (Condo and Bishop 1995)

Through this document and through its Web site, the U.S. MAB Program continued to highlight ISDA as one of several "partnerships" that "have developed at the regional level among groups interested in participation in the biosphere reserve principles of conservation of biological diversity and development of environmentally compatible economic use" (U.S. Department of State 1998).

Thus, by 1995, ISDA was nominally a "biosphere reserve cooperative" —or, as Tony Ramon and Superintendent Smith reported in October to a U.S. MAB Program conference in Washington, D.C., ISDA was the core of *the* "Sonoran Desert Biosphere Cooperative" (Condo 1996). Tellingly, a now defunct NPS Web page described ISDA as a MAB Cooperative "without MAB biosphere reserve designation" (U.S. National Park Service 1998). But even if it was not apparent at the time—and more than likely it was—the long-term survival of a broad-based biosphere reserve cooperative was unlikely to succeed without the congruent establishment of a geographically encompassing biosphere reserve.

An International Biosphere Reserve for the Sonoran Desert?

By June 1994, the Taskforce had prepared a draft proposal for an international Sonoran Desert biosphere reserve. It is worth pausing here to note that efforts to integrate land management in the Sonoran Desert had a storied history of efforts to integrate ORPI and Cabeza Prieta as the "Sonoran Desert National Park" (U.S. National Park Service 1965). Having hiked extensively in the area while practicing law in Tucson, former Secretary of the Interior Stewart Udall (1961–69) would later "claim some credit" for the idea: "What I was proposing to President Johnson was seven million acres; that would have been a record, in terms of acreage, and I just said it right out for him, 'Mr. President, if four million acres was just about right for Herbert Hoover, seven million is right for Lyndon Johnson'" (Udall 1997, 316, 318). Conservation organizations such as the Sierra Club and influential decision makers such as Congressman Morris Udall (D-AZ, and the Interior Secretary's brother) took up the banner for such a park, but it is unclear how much energy was put into the effort (Udall 1966). The former Secretary later recalled that the idea "didn't receive a lot of publicity" and that he could not convince President Lyndon B. Johnson to do it (Udall 1997, 317).

Although Johnson ignored Udall's proposal, the seed of an idea had been planted. Significantly, the proposal issued by the NPS had stated, "It can only be concluded that the entire area is eminently qualified for National—if not International—Park status" (U.S. National Park Service 1965, 29). The Taskforce was now, in a sense, following up by planning to present its "bold plan" to the Secretary of the Interior and the MAB Program at some point before September 1994. However, it appears that the publicity-based "major campaign" was put on hold for reasons that are now difficult to discern (see Nabhan et al. n.d.). What appears to have happened is that Nabhan presented the idea to the ISDA board in August, which led to "the board's decision to support the idea and to begin a series of community workshops to gain greater local input on the proposal" (ISDA Board of Directors 1994; TPSDBR 1994). ISDA went ahead with the workshops, but, as will become apparent, the Taskforce was mistaken if it was interpreting the "support" from ISDA as unconditional.

The Taskforce then set in motion a range of activities, including field trips to the Sonoran Desert for high-level bureaucrats from the U.S. National MAB Committee, letters and phone calls to influential individuals (including Stewart Udall), extensive preparation for a press release on the effort, and preparation for an opinion survey (Camp 1994; Nabhan et al. n.d.; TPSDBR 1994). The Taskforce reported "a great deal of support from many places," including at least one letter of support from a U.S. agency (Pulliam 1994), as well as a letter of support from the Northern Regional Forum of the National Project on the Protected Natural Areas of Mexico (Gómez-Pompa et al. 1994). At this point, Nabhan was concerned but optimistic over the effort's prospects, noting that he anticipated "an administrative ratification of a regional man and the biosphere cooperative program and expanded reserve within the next six months." Warning that "we need to carefully read political changes in Washington," he also believed that "community sign-offs, not administrative obstacles, are where [we] need to focus" (Nabhan 1994).

Then, at some point around the beginning of 1995, Nabhan produced a document entitled "Completion of the Sonoran Desert Biosphere Network along the U.S./Mexico Border." Reviewed by several ISDA participants, including Laird, Nagel, and ORPI Superintendent Smith, the document stated,

> ISDA has brought together U.S., Mexican, O'odham, and Cocopah spokespersons to explore opportunities for enhanced cooperation within this larger ecosystem. . . . A tremendous amount of cooperation

has been generated from ISDA conferences and meetings. This [led] one Nature Conservancy administrator to suggest that ISDA is already informally serving many of the same functions that a Regional Man and the Biosphere Cooperative Program would serve. (Nabhan 1995)

Nabhan presented the document to the ISDA board, explaining that it had been "hastily prepared" for a UNESCO Intergovernmental Conference on Biosphere Reserves to be held in Seville, Spain, in March 1995 (see Batisse 1997). ISDA participants who would then travel to that conference included Nagel, ORPI superintendent Smith, and two members of the Tohono O'odham Nation: Floyd Flores and Fernando Valentine. Of the hundreds of biosphere reserve groups represented at the conference, Nagel told me, ISDA's was the only one that had indigenous people in attendance. He reminisced that after their presentation that culminated in a celebration of the "spiritual dimensions of the biosphere reserve," they received a standing ovation.

Thus, around the beginning of 1995, everything seemed to be falling into place: The Taskforce was holding several influential political levers, Nabhan had a nascent draft biosphere reserve proposal in hand, and ISDA seemed poised, as Nabhan put it, to be the "community sign-off" for a Sonoran Desert Biosphere Reserve. Yet at this point, the effort hit a roadblock. Apparently, there was still dissension within ISDA over the degree to which a biosphere reserve approach—or at least, a biosphere reserve label—was appropriate for the western Sonoran Desert. The conflict came to a head upon the return of those ISDA participants who had attended the Seville Biosphere Reserve conference. Despite the favorable reception the attendees had received in Seville, some ISDA board members accused the attendees of representing ISDA without prior approval.

Notably, it was just at this time that Laird had decided to move on, and ISDA had hired a new coordinator, Ceal Smith. Smith held the position for only about nine months, her tenure abruptly coming to an end largely due to an acrimonious conflict with a particular board member (ISDA 1995b). Looking back a half decade later, some ISDA participants—including Smith herself—described her experience as one of being dropped down into the middle of competing forces within ISDA. From Smith's perspective, one side was the "grassroots board," the other a number of strong proponents of the biosphere reserve proposal. A major part of the problem, she further argued, was that much of the debate took place outside of ISDA and at the meetings of the Taskforce. Others, however, saw it much differently. Laird, for instance, argued that ISDA at that point was simply too complex an organization to manage for anybody without a solid institutional memory—something that Smith lacked through no fault of her own.

In short, internal tensions over both staffing issues and the biosphere reserve designation were running high and would end up having serious consequences. Nonetheless, after meetings between ISDA's board and Nabhan, the board ultimately voted for the biosphere reserve proposal. Subsequently, over the course of 1995 and into 1996, Nabhan worked with ISDA participants to update and expand the proposal for submission to the U.S. MAB directorate. While his recollections about the development of this document indicate that it may have been more influenced by the Taskforce, the report credited authorship as follows: "Prepared by Gary Nabhan, Arizona–Sonora Desert Museum, with review and comments by agencies and by board members of the International Sonoran Desert Alliance" (Nabhan 1996). Not dramatically changed from the first version drafted for Seville, the January 1996 draft stated that across no other international border "has there been a better opportunity to establish collaborative management . . . at the landscape—or ecosystem-scale," and that an expanded Sonoran Desert biosphere reserve would meet "many of the U.S. MAB criteria that Organ Pipe Cactus National Monument, in and of itself, does not." The report further argued that,

> Perhaps the greatest benefit of reserve expansion would be to allow landscape- or ecosystem-level management of threatened organisms which are now suffering from the disruption of corridors and ecological processes between areas. . . . An overwhelming majority of the region's land managers and conservation biologists surveyed believe that the ten most vulnerable species in the area would be better served by expanding MAB-style management across administrative and international boundaries. (Nabhan 1996, 6–7)

Once again, hopes seemed to be running high. A full proposal was on the table; ISDA had moved past its staffing issues (Laird had returned to replace Smith); and while internal division within the board over biosphere reserve designation had unlikely disappeared, it had been democratically resolved. But just around the time that the second report was completed, according to Nabhan, "we started to get mixed messages."

Although MAB officials in Washington, D.C., were still supportive, NPS officials "began to say that we were going too fast, and that it was unclear whether all the agencies were lined up with one another."

Thus, having overcome serious internal challenges, the effort to create an international Sonoran Desert biosphere reserve suddenly found itself marching toward a closing door—a political and bureaucratic closing door that was external to the region they were trying to protect. Those "mixed messages" Nabhan was receiving were the first indications of the imminent demise of the biosphere reserve proposal.

The Biosphere Reserve as Political Pariah

By 1995, ISDA had been featured in several regional periodicals and reports, receiving high praise for its efforts to establish biosphere reserves and MAB concepts in the Sonoran Desert and garnering a reputation of effectiveness and credibility (ISDA 1994a; Planeta.com 1995; Steffens 1994; Vesilind 1994; Williams 1994). ISDA's leadership also saw its work as having successfully implemented a biosphere reserve cooperative through which ISDA had "succeeded in building self confidence and self-reliance among its members" and had generated "a regional identity and a common understanding and support for the biosphere reserve program" (Condo and Bishop 1995; see also Laird 1994b; Laird et al. 1997).

Such enthusiasm bolstered hopes for an international biosphere reserve designation; indeed, the effort was effective enough to allow some to imply that the reserve had actually been created (see, for example, Nimkin n.d.). Yet no international Sonoran Desert biosphere reserve was ever established. Given how close its proponents seem to have come, the absence of such a designation raises two intricately woven questions:

- Why is there now no international Sonoran Desert biosphere reserve?
- Was the effort for naught—that is, was ISDA ultimately ineffective in this regard?

A concise, integrated answer to both questions is that a combination of internal problems and external political forces deflated the international biosphere reserve proposal into a lower profile effort to expand existing cooperative activities in the region. But this short answer is hardly satisfactory, and the long answer requires addressing each question separately.

The Demise of the International Biosphere Reserve Effort

Looking back at that time period a half decade on, many of those involved in the effort had difficulty articulating why the initiative failed to establish a biosphere reserve in the region. Some variant of "it just was not the right time" was a common refrain. Undoubtedly, my queries met with such ambiguity at least partly because they dredged up unpleasant memories. The tenure and termination of Ceal Smith's employment engendered a bitter antagonism within ISDA's leadership and apparently among some board members. Smith made two principal charges against the way ISDA was structured. First, she believed that the board never adequately represented the three main cultural groups in the area and thus was never able to attain any real decision-making power. Second, she accused outside conservationists of cynically using ISDA as a grassroots front for promoting their agenda—as well as for fundraising. To the first charge, Laird countered that the "internal chaos" within ISDA was the result of Smith's lack of leadership experience, which Laird believed had rendered the board into factions that worked poorly with each other. To the second charge, Laird argued that SI had been scraping by to support ISDA for nearly five years.

Sorting out these two perspectives is difficult. Both are true to a degree, yet neither gets to the essence of what happened within ISDA. In response to Smith's first charge, although it is true that ISDA was not fully representative of the region, it had come far closer to the ideal of diverse representation than any other effort. Indeed, one of the reasons it moved so slowly was all the logistical problems brought on by allowing for real cultural and linguistic diversity. To Smith's second charge, although outside entities undoubtedly rode the coattails of ISDA, there is no indication that funding granted to ISDA was ever significantly diverted from its programmatic and outreach activities. More specifically, the relationship between SI and ISDA appears never to have been anything worse than reciprocal—particularly in light of Laird's nine months of volunteer work and countless hours of unpaid overtime.

In response to Laird, even though Smith may not have been the right person at the right time for ISDA, her presence alone would have been unlikely to engender such divisiveness within the board had there not been a fundamental difference of belief lurking below the surface. Indeed, it should be recalled that the controversy over the fact that some ISDA participants (and not others) had attended the Seville Conference on Biosphere Reserves occurred just as Smith was arriving on the scene.

Confidentiality limits what can be told of the rest of this story, but it amounts to little more than yet another iteration of the divisive conflicts that often arise in institutional settings. More important than a storyline of personal conflict, however, is the fact that much of the internal divisiveness resulted from the incessant barrage of practical operational challenges that ISDA faced—including geographical isolation, the necessity of translating documents, a lack of communication technologies (few fax machines, not to mention computers or modems), and even a "bad mail system" (Keystone National Policy Dialogue on Ecosystem Management 1995). Place all of these basic communications problems on top of the intangible boundaries between extremely diverse cultures—and what you had was a situation highly combustible and bogged down by inertia.

Yet despite these conflicts internal to ISDA, the demise of the international biosphere reserve proposal can hardly be ascribed to internecine battles alone. Indeed, despite her reservations, under Smith the group had evolved into an active partner in promoting an international biosphere reserve for the Sonoran Desert. So even had ISDA been fully in accord within itself, external negative pressures would likely have proven insurmountable. And these external pressures were many. Nabhan, for instance, cited the cultural insensitivity and high staff turnover of federal land managers as a root cause of the problems facing the biosphere reserve proposals. Laird pointed to a number of other specific challenges: The national attention NAFTA had brought to the border region was waning, federal U.S. agencies "could not agree on how to cooperate and participate for fear of losing jurisdiction," and there were frustrations with the constant delay in management plans for the Pinacate and Upper Gulf Biosphere Reserves. Furthermore, there were several other bilateral initiatives on the table—including the creation of an economic thoroughfare through Organ Pipe to Puerto Peñasco—over which "there was no agreement, particularly community agreement." Finally, Laird perceived declining interest in biosphere reserves due to a lack of understanding of what the status would actually do for the area. Given all of these impediments, Laird's general assessment was that the effort "kind of just burst."

There was, however, another factor, one alluded to in several of my interviews, that overlaid all others in the demise of the international biosphere reserve. Nagel articulated this factor as "not something specific that put it off the agenda," but as a pervasive governmental mistrust in bilateral cooperation over the countries' shared border. "Creating international 'anythings' in the border area is not popular with government agencies," he noted. "If something threatens the sovereignty of the international boundary, it will be viewed with suspicion by government agencies on both sides." This had certainly been the case for Mexico, which, as Ezcurra described, historically interpreted suggestions of "border peace parks" as little more than a U.S. grab for greater control of the region. But whereas Mexico had gradually become more open to the idea of border biosphere reserves, the United States was becoming more hesitant.

The hesitation resulted from political forces that were coalescing to extinguish any chance of success in biosphere reserve designation—particularly any designation with the added appellation of "international." ISDA's efforts to promote the biosphere reserve concept were becoming politically viable just as there was a spike in vocal opposition to international conservation initiatives, opposition that was aimed not only at the MAB Program but also at U.S. cooperation in the World Heritage Convention and the Biodiversity Convention. As Rikoon and Goedeke (2000, 2, 5) have described, this opposition remains a complex phenomenon that arose from an "extremely heterogeneous" movement defying simplistic critiques from the environmental community that such "anti-environmentalism" constitutes nothing but a "corporate-led reactionary struggle incorporating a relatively small number of people and interest groups engaged in disinformation and acts of anti-environmental violence." Such rhetoric serves not only "to underestimate the number of its participants and sympathizers," but also "ignores the increasing examples of grassroots opposition and concerns" (Rikoon and Goedeke 2000, 4, 9). Although this is an important critique, much of the antagonism toward biosphere reserves nonetheless did emanate out of several politically conservative (and, arguably, homogenous) institutions, arousing fears that U.S. participation in international conservation programs would inevitably lead to a loss of territorial sovereignty. The antagonism was adopted both by the "wise use" movement, which had grown into a politically powerful entity after the 1994 Republican capture of both chambers of the U.S. Congress, and—at its most extreme—by the "American militia movement," which had credulously associated mysterious over-flights of "black helicopters" with an impending invasion by blue-helmeted UN troops (Rowell 1996; Stern 1996).

The antagonism may have reached its peak in 1995 when UNESCO designated Yellowstone National Park as a "park in danger" (McHugh 2000, Turner 2002). Suddenly, biosphere reserves and world heritage sites became loaded topics engendering vociferous hostility toward anything related to the United Nations. But these fears were, for the better part, patently unfounded (if not, in the more extreme cases, laughable). The U.S. Congressional Research Service, for instance, pointed out that a country's participation in the MAB Program is entirely voluntary, the program imposes legally binding obligations, and the International MAB Programme explicitly recognizes each country's sole sovereignty over its biosphere reserves (de Klemm and Shine 1993; Fletcher 1999; Soles 1998). Of course, loss of sovereignty is not the only debating point; opponents have also argued that the process of designating biosphere reserves constitutes an unreasonable delegation of Congressional authority, that biosphere reserves foster a radical environmental agenda, that they provide conservationists with undue influence, and that they simply amount to a waste of taxpayer resources. Conservationists, it should come as no surprise, have ready replies to all of these arguments. But wherever one lands on these issues, in terms of the most contentious point, biosphere reserves specifically and overtly do not entail loss of U.S. sovereignty.

Regardless of this fact, however, distrust of international cooperative conservation initiatives led to significant repercussions at both local and federal levels. While Arizona does not appear to have experienced the level of strident local opposition to biosphere reserve proposals as other areas of the country—such as the Ozark Man and the Biosphere Reserve (Rikoon and Goedeke 2000) or the Adirondack-Champlain Biosphere Reserve (Helvarg 1994, 224–225)—several ISDA participants noted their frustration with the "strong anti-UN, anti–one world kind of thinking" in the state (as Ceal Smith's successor Reggie Cantú put it). ISDA board member and ORPI ranger Dominic Cardea noted that "paranoia about the UN" and biosphere reserves had some parks taking "so much flak" that they were considering removing the Man and the Biosphere plaques. "You'd be surprised," he noted, "how many people come to the visitor's center and tell us that they'd see black helicopters that didn't have any markings and they can't see anything in it—but they could see that it did have guys with blue helmets in there." Notably, both Laird and SI staffer Steve Cornelius pointed out that within the two other main cultures of ISDA—O'odham and Mexican—biosphere reserves aroused nothing comparable to the paranoia encountered in the United States. Whereas the biosphere reserve model worked "very well" in the "schema" of the Tohono O'odham and was one of "the better funded, administered entities" in Mexico, Laird recalled that on the U.S. side of the border "the Ajo community particularly didn't like or understand it."

At the federal level, by 1995 Representative Don Young (R-AK) had introduced the American Lands Sovereignty Act (ALSA), which would have put severe restrictions on the designation of biosphere reserves (Lamb 1997; U.S. Congress n.d.). The House passed the Act in 1997 after contentious congressional hearings (U.S. Congress n.d.). Although a companion bill failed to pass in the Senate, the Clinton administration was hardly immune to these substantial political pressures and consequently became extremely wary of setting U.S. public lands policy in the context of international affairs. With biosphere reserves having achieved "pariah status," as SI's Cornelius put it, the administration "carefully distanced itself" from "anything that might remotely resemble black helicopters"— including the proposal for an international Sonoran Desert biosphere reserve (Zakin 1997).

As Nabhan summed up, they had pitched the international biosphere proposal "at the most politically inopportune time that we could." That the biosphere reserve proposal was going nowhere was becoming apparent by the end of 1996. But even as hopes were evaporating, a new window of opportunity was opening for international cooperation in the western Sonoran Desert.

Agreements to Agree: The Arizona-Sonora MOU and the U.S.-Mexico Letter of Intent

In November 1996, Arizona governor Fife Symington and Sonora governor Manlio Fabio Beltrones signed a Memorandum of Understanding (MOU) announcing a joint endorsement of a Binational Network of Sonoran Desert Biosphere Reserves (Inter-Hemispheric Resource Center 1997; Pearson 1998, 13). Incorporating most of the public lands in the western Sonoran Desert, the network was to promote "an integrated program which protects cultural values, promotes sustainable community and economic development in the region, and promotes cooperation between the contiguous protected areas on both sides of the border so as to motivate collaborative resource management of the region's shared resources" (U.S. Department of State 1997).

ISDA's leadership, ISDA participants, the press, and the U.S. National Committee for the MAB Program have all credited ISDA for bringing about this intergovernmental initiative, which was described as the potential "beginning of a binationally managed biosphere reserve" and "an unprecedented event in international environmental cooperation" (Inter-Hemispheric Resource Center 1997; Muro 1997; Nagel and Cantu 2001). However, although Laird recalled that ISDA was "right in the middle" of the overall effort to get the MOU signed, she was unwilling to draw a causal connection from ISDA to the MOU. When I asked her if the MOU would have occurred without ISDA, she responded that it "might have" and pointed out that much of the impetus came from Nabhan's Taskforce. Furthermore, both Cornelius and Nagel pointed out that Propst's political connections were crucial in establishing the MOU, and that support from the Sonoran secretary for the environment, Maria Elena Barajas, was also instrumental.

Regardless of how it came about, the Symington-Beltrones MOU seemed to reinvigorate enthusiasm for the idea of an international biosphere reserve. In December 1996, ORPI superintendent Harold Smith and Cabeza Prieta NWR manager Bob Schumacher wrote a letter to the MAB Biosphere Reserve Committee "to encourage all involved to keep moving ahead with the progress toward a transboundary Sonoran Desert Biosphere Reserve Network." Describing their lands as a "defacto biosphere," the two land managers recommended that (1) the biosphere reserve designation for ORPI be expanded to Cabeza Prieta NWR, (2) ISDA be recognized as the "equivalent of a Man and the Biosphere Cooperative Program," and (3) U.S. land management officials increase cooperation with the adjacent Mexican biosphere reserves (Smith and Schumacher 1996).

Although the documentary record is not complete and people's memories are unclear on the chronology, the Smith-Schumacher letter appears to be the last real push to designate the area as a biosphere reserve. By early 1997, the effort to create an international Sonoran Desert biosphere reserve was widely recognized as politically infeasible, and proponents for international cooperation were relying on the MOU's label of "binational network." For example, in February of that year a group of land managers, community leaders, and NGOs—including ISDA, SI, and Nabhan's Arizona–Sonora Desert Museum—wrote a letter entitled "Toward completion of a transboundary Sonoran Desert Network of Protected Areas, with multi-cultural community involvement" (Transboundary Sonora Desert Letter 1997). The authors echo the sentiments of the Symington-Beltrones MOU and note that ISDA "and its affiliates on both sides of the border have held community workshops and engaged in multi-year discussions of community involvement in the management of public lands" (Transboundary Sonora Desert Letter 1997).

One month later, on March 22, 1997, ISDA's board met on the border —literally *on* the border—"to dramatize a host of difficulties caused by tightened border security" (Zakin 1997). Zakin described the scene as looking "like a Jewish Orthodox bar mitzvah: American members sat on the U.S. side of the fence; Mexican members sat on the other." ISDA's hope was that the stunt would help push DOI secretary Bruce Babbitt and Mexico's secretary of the environment Julia Carabias to come to an agreement regarding "better border cooperation"—an agreement that might even be taken up by U.S. President Bill Clinton and Mexican president Ernesto Zedillo at their upcoming May 1997 meeting in Mexico City (Zakin 1997).

That hope, unlike the hope for an international biosphere reserve, was about to become a reality. On May 5, 1997, two months after ISDA's on-the-border meeting and just before the Zedillo-Clinton summit, Secretaries Babbitt and Carabias signed a Letter of Intent (LOI) in Mexico City (Carabias and Babbitt 1997; Pearson 1998, 13; Sonoran Institute n.d.). The purpose of the LOI was to "expand existing cooperative activities in the conservation of contiguous natural protected areas in the border zone, and to consider new opportunities for cooperation in the protection of natural protected areas." It referred to two specific pilot projects, one for the western Sonoran Desert, the other for the Big Bend region bordering the states of Texas, Coahuila, and Chihuahua (Carabias and Babbitt 1997; Cisneros and Naylor 1999; Laird et al. 1997). Although not under a biosphere reserve agreement, both federal governments were now at least nominally working together in the western Sonoran Desert under the title of "sister areas" (U.S.-Mexico Border Field Coordinating Committee 2001).

Could ISDA look upon the LOI and its sister areas as its own success? That is, did ISDA play a significant role in establishing the LOI? Once again, the causal trail is unclear. While the historical record confirms that ISDA generated support for the LOI, there is no hard evidence that such support was a major factor leading up to the LOI. Susan Zakin (1997), who may have been the only journalist following the story at the time, implied that although ISDA was aware of the high-level negotiations between Mexico and the United States, it did not have a direct influence on them. Furthermore, several other factors seem to have played an equal or greater role. First, the LOI itself directly ascribes its origins to a series of high-level U.S.-Mexico agreements and not to any regional efforts (Carabias and Babbitt 1997). Second, the state-level MOU seems to have generated its own political momentum toward a federal-level agreement.

While some of that momentum may have consisted of enthusiasm for the idea of transborder cooperation, much of it undoubtedly resulted from political maneuvering at the upper echelons. In particular, having voiced its displeasure at the state of Sonora for signing the MOU with the state of Arizona "without prior federal approval," the Mexican federal government could well have been using the LOI as a way of reining in the waywardly independent state (Inter-Hemispheric Resource Center 1997).

Third, and perhaps most important, after the NPS hosted a delegation of officials from Mexico's Ministry of Environment and Natural Resources for observation and study of the Waterton-Glacier International Peace Park between Canada and the United States in 1996, Mexico had proffered a specific "diplomatic initiative" to the United States in early 1997 regarding the establishment of binational protected areas (Zakin 1997). This initiative dated back to efforts beginning in the 1930s to create a peace park between Big Bend National Park in Texas and what would become the Maderas del Carmen Protected Area in Coahuila and the Cañon de Santa Elena Protected Area in Chihuahua (see Sifford and Chester, forthcoming; Cisneros and Naylor 1999). The 1997 Mexican proposal was "received with some surprise by U.S. Department of the Interior and National Park Service officials," who apparently had not expected such an expeditious response to the Waterton-Glacier visit (Cisneros and Naylor 1999). While the proposal focused on the Big Bend region, it also referenced both the Pinacate and Upper Gulf Biosphere Reserves. In its positive response to the Mexican proposal, the DOI nonetheless deleted "all reference to the formulation of a bilateral legal instrument to regulate the establishment of binational protected areas," and ultimately the LOI would not refer to binational protected areas—and certainly not to anything regarding international biosphere reserves (Cisneros and Naylor 1999).

A fourth possible factor leading up to the LOI was another—one could at this point say *yet* another—cooperative initiative within the Sonoran Desert. At the Eighth Annual U.S.-Mexico Border States Conference on Recreation, Parks and Wildlife in February 1997, an ad hoc gathering of land managers and NGO representatives came to a general consensus to form a Sonoran Desert Ecosystem Council (Parra-Salazar et al. 1997). At a follow-up meeting soon after in Phoenix, the agency representatives pulled back from the word "council" (which Steve Cornelius told me, "implied another level of bureaucracy") in favor of "partnership." The Sonoran Desert Ecosystem Partnership (SDEP) was then formally agreed to at the LOI's first work plan session in Puerto Peñasco in October 1997, which was attended by representatives from over twenty Mexican and U.S. federal and state management agencies, universities, tribal groups, and NGOs (Cornelius 1998, Table 3; Pearson 1998, 13; U.S.-Mexico Border Field Coordinating Committee 2001). Cornelius and ISDA's board president Carlos Yruretagoyena volunteered to "facilitate a process to define the operational function of a Sonoran Desert Ecosystem Partnership" among the disparate groups. The governmental officials accepted the offer, and over the course of the following year Cornelius and Yruretagoyena held a series of meetings to define the SDEP's geographic scope, membership, financing, authority, and mission. The official presentation of the SDEP occurred in Tucson in June 1998 at the Ninth U.S.-Mexico Border States Conference on Recreation, Parks and Wildlife. Cornelius pointed out that although they were developing in tandem, the SDEP and the LOI had been separate processes that were not unified until July 1999.

To recap, the LOI was highly influenced by the four factors of (1) previous federal-level agreements, (2) political momentum coming out of the state-level MOU, (3) a Mexican initiative focused on the Big Bend area in Texas, and (4) the SDEP process. In light of these multiple causal factors, how likely was it that ISDA had a role in creating the LOI? Ultimately, the few individuals I spoke with who were familiar with this intricate bureaucratic history came to the same broad conclusion: Although an LOI would have come about with or without ISDA, ISDA had generated widespread interest in the western Sonoran Desert and had demonstrated local support for international cooperation—and in so doing had played an important role in positioning the western Sonoran Desert to be chosen as one of the two pilot areas under the LOI.

Transborder Cooperation without the Biosphere Label

Although tensions within ISDA hampered the effort to establish an international Sonoran Desert biosphere reserve, the demise of the initiative mostly provides a harsh demonstration of how larger societal and political forces can vitiate years of hard work. Surprisingly, in my talks with many of the actors involved in ISDA during that time period, few bemoaned the demise of the effort. Although they did not avoid the issue, most of them demonstrated a resolve not to focus on why or how the effort dissipated. Rather, they looked to ISDA's significant role in (1) influencing the Mexican government to designate two biosphere reserves in the area, (2) coordinating the actions of a number of nongovernmental and federal actors who otherwise probably would never have met, and (3) promoting new and more widespread interest in conservation and cooperation across geographic and cultural borders.

Some ISDA participants argued that the effort never truly disappeared but simply dropped the politically charged appellation of "biosphere reserve." For example, Laird et al. (1997, 308) argued that whether or not the proposed biosphere reserve network were to be formally adopted, "it is essentially in place, occurring at an informal level among local community leaders, resource managers, conservation groups, and business owners who have fostered an innovative alliance to protect the desert." Propst had similar recollections, and because of his position "halfway" between ISDA and many of the political actors with considerable influence over land management in the Sonoran Desert, it is worth quoting him at length:

> The whole MAB Program became a lightning rod, and we realized locally we didn't need the MAB designation to do what we were doing. The lightning rod was coming from Washington and from the right wing—property rights groups and so forth. And our feeling was if the right wing wants to attack the MAB concept, fine—it's not hurting anything. And then as a result of the right-wing attacks, the whole biosphere reserve program within MAB dissipated. They just kind of quit funding it, and MAB retreated back to its hole of being just kind of a forum for scientists. And so we dropped it; why fight a battle over surrogate issues?
>
> And we continued to do what MAB's all about. It's really ironic because the right wing has attacked the biosphere reserve concept when, if you presented it to them with different words and without the UN thing, it's just what they want—they want to recognize the importance of local folks. But it became this sovereignty thing. And I was thinking: Why do we want to weigh this whole effort down with some symbolic fight with Helen Chenoweth [a U.S. Congressman —as she insists on being called—noted for her legislative efforts to disband U.S. environmental laws] that just doesn't matter? So we just figured that the best thing to do is to do it without reference to the name that seems to be a lightning rod.
>
> You know, ideally the biosphere reserve program would take hold.

But we've just accepted the fact that within the U.S. it won't. And what's happened is that all over the U.S. we're seeing the concepts applied, but without a central framework. And that's probably okay; I'd rather see it happen organically bottom up, a thousand flowers bloom, than to create some kind of a United States Institute for Creating . . . [hand-waving gesture]

Both Cornelius and Nabhan concurred. The SDEP, as Cornelius argued, entailed "all the principles of a biosphere reserve without calling it a biosphere reserve and without structuring or establishing a new structure. . . . This is actually no different. It's the same thing, they're just talking about a network of protected areas." Nabhan pointed out that after the failed attempt for a biosphere designation, the Sonoran Desert Taskforce was "promised" by the DOI International Affairs Office that the SDEP "would enable most of the activities that we wanted to do through a paired biosphere reserve system anyway." Given the creation of both the LOI and SDEP, the government seems to have at least nominally kept its promise.

In sum, ISDA—including the entire process of putting ISDA together—did significantly alter the status quo of land management in the Sonoran Desert. Despite the unique character of each biosphere reserve experience, whether it be in the savannas of Africa, the mountains of Asia, or the forests of Latin America, the lesson from the Sonoran Desert is clear and transferable: In contemplating whether to engage in "transborder conservation," do not be intimidated by the threat of failure or fact of ominous societal forces working against the prospect of success. The Sonoran Desert would be a different and poorer place had such intimidation prevailed.

ISDA Today and Tomorrow

After 1997, ISDA's efforts to create an international biosphere reserve evaporated, and ISDA adopted a significantly lower profile in land management issues in general. Numerous ISDA participants recalled to me that organizationally it was an extremely difficult period for ISDA; it took some time, Nagel said, for it to recover from the dual shocks of its own internal staff-board conflicts and the assault on anything related to biosphere reserves.

It is worth noting in this context that ISDA did not substantially participate in two post-1997 high-profile conservation efforts in the region. The first, led by Tucson-based activist Bill Broyles, was to revive Stewart Udall's idea of combining ORPI, the Cabeza Prieta NWR, and other federal lands into a unified Sonoran Desert national park (SDNP) and ultimately a U.S.-Mexico peace park (Sonoran Desert National Park Project 2001). Although Broyles would partially attribute the reinvigorated campaign to ISDA and the 1992 Land Use Forum (in which he had participated), he also argued that the "ISDA ideal" had been "unable to move custom or inertia." The SDNP and peace park initiatives have yet to make significant headway (Sifford and Chester, forthcoming).

The second effort, also with Broyles as one of the lead activists, was to carve 2,009 square kilometers (776 square miles) out of BMGR and BLM lands to create the Sonoran Desert National Monument (SDNM) south of Phoenix, an effort that was successfully accomplished during the last days of the Clinton administration in January 2001. Notably, ISDA was not among the nineteen institutions supporting the effort (Felger et al., forthcoming).

There are different interpretations of why ISDA did not participate in these two initiatives. Several ISDA participants argued that both were simply outside of ISDA's domain. First, the SDNP effort has been led by a number of dedicated—some would say zealous—conservationists in the region and not by a consensus-driven organization such as ISDA. Second, the SDNM is located just south of Phoenix and thus on the far edge of ISDA's traditional geographic coverage. Thus, even if ISDA had successfully established an international Sonoran Desert biosphere reserve, one should not assume that ISDA would necessarily have participated in these two subsequent efforts. Nevertheless, I find it difficult to believe that had ISDA not undergone so many travails, it would not at least have become one of the many stakeholders in those two efforts. Such conjecture is, however, my own hypothetical alternative scenario, one that is not shared with several knowledgeable individuals working in the Sonoran Desert.

After 1997, as part of a plan to strengthen itself organizationally, ISDA shifted its programmatic emphasis away from land management issues toward the realm of environmental education and community outreach, sponsoring numerous place-based programs with funding principally coming from the Ford Foundation. Around 1999, it became involved in an effort to remodel the Curley School complex in Ajo. Built in 1919 in the Spanish colonial-revival style, the school's domed clock tower dominates the town plaza as a recognizable landmark for locals and tourists alike. The complex has largely stood abandoned for the past two decades, with ISDA and its Regional Resource Center being the sole occupants for much of the time since moving into the building. Current executive director of ISDA Tracy Taft has put ISDA's weight behind the Curley School effort. As a result, if one were to have stepped into an ISDA board meeting in 2005, one would have heard far more about architectural plans and market research than conservation or cultural preservation. The project is ambitious, incorporating low-income housing and working space for artists and their families, community space for artists and Ajo residents, commercial and office space for arts-related businesses, and space for nonprofit organizations, including ISDA. In relation to this project, ISDA has also taken a leadership role in the Ajo School Arts Partnership, which brings community artists into the Ajo schools (Tracy Taft, pers. comm. December 2004).

Engaged as she is in the Curley School project, Taft does not intend to permanently reformulate ISDA into a local booster organization. She believes that the Curley School project will revitalize Ajo, which has never taken advantage of being the "gateway town" to multiple public lands and the Mexican tourism destination of Puerto Peñasco. "Nearly everybody who travels through Ajo stops for gas," she noted, "but the Curley School project will make them stop to see Ajo." When the Curley School project is completed, Taft believes that ISDA will be integrally associated with it, which will put ISDA back on the map for both local residents and local land managers—managers whose jobs are directly affected by the economic success (and failure) of the town of Ajo.

At that point, Taft argues, ISDA will be able to refocus on land management issues. And to a large degree, she has already made inroads in reestablishing ISDA's ties to the land management agencies by taking a lead role in establishing a network of gardens as part of a regional effort to protect "nectar corridors," which have been scientifically defined as "latitudinally broad paths of blooming plants pollinated primarily by migrant nectar-feeders" such as bats and butterflies as they make their way between winter and summer habitats (Fleming 1994). Four land management agencies are involved in the project, and ORPI has tied their work to protect the Quitobaquito pupfish (*Cyprinodon macularius eremus*) with the nectar gardens. Taft has also been actively involved in the BLM's five-year land-use planning process, particularly in regard to recruiting participation from Ajo residents and grappling with the issue of off-road vehicle use.

In the meantime, Taft has been organizationally restructuring ISDA's board and overhauling its financial system. There is no guarantee, of course, that all this groundwork will lead to success in the next few years—particularly since the substantive and procedural challenges facing ISDA have only expanded since the early 1990s. But with Ajo standing at the edge of what could be an economic revitalization—and all of the benefits and problems such a revitalization would inevitably bring to the broader western Sonoran Desert—ISDA is poised to become once again a voice for its original core mission: to promote a "healthy, positive relationship between the Sonoran Desert and its inhabitants, and the needs of humanity."

References

Abbott, Carl. 1993. *The metropolitan frontier: Cities in the modern American West.* Tucson: University of Arizona Press.

Ajo Copper News. 1992. WPCCC told of planned land forum. April 22, 1.

Ajo Copper News. 2000. Deadly desert does little to stem tide of immigrants. September 13, 6.

Ajo Copper News. 2001. Area agencies must deal with heavy UDA trafffic. April 25, 6.

Alaimo, Carol Ann, and Joseph Barios. 2000. Teen mom dies to save tot. *Arizona Daily Star*, June 1, A1.

Allen, Paula Gunn. 1997. Cuentos de la tierra encantada: Magic and realism in the southwest borderlands. In *Many wests: Place, culture, and regional identity*, ed. David M. Wrobel and Michael Steiner, 342–365. Lawrence: University Press of Kansas.

Anonymous. 1994a. *Biosphere Dates.* Puerto Peñasco, Sonora. January. Files of the International Sonoran Desert Alliance, Ajo, AZ.

Anonymous. 1994b. Convenio de Colaboración entre Dependencias para Establecer y Operar una Cooperativa del Hombre en la Biosfera en el Desierto Sonorense. 1994. Files of the International Sonoran Desert Alliance, Ajo, AZ.

Arizona Department of Commerce. n.d. *Sells/Tohono O'odham Reservation community profile.* http://www.commerce.state.az.us/pdf/commasst/comm/sells.pdf.

Batisse, Michel. 1997. Biosphere reserves: A challenge for biodiversity conservation and regional development. *Environment* (June): 6–15, 31–33.

Beck, Warren A., and Ynez D. Haase. 1989. *Historical atlas of the American West.* Norman: University of Oklahoma Press.

Bennett, Peter S. 1989. Preface. In *Report on treaties, agreements, and accords affecting natural resource management at Organ Pipe Cactus National Monument (1988)*, ed. Carlos Nagel. Special Report No. 8. Tucson: University of Arizona, Cooperative National Park Resources Studies Unit.

Bowden, Charles. 1977. *Killing the hidden waters.* Austin: University of Texas Press.

Broyles, Bill. 1992. Letter to Lincoln Institute, November 16. Files of the International Sonoran Desert Alliance, Ajo, AZ.

Broyles, Bill, Richard S. Felger, Gary Paul Nabhan, and Luke Evans. 1997. Our grand desert: A gazetteer for northwestern Sonora, southwestern Arizona, and northeastern Baja California. *Journal of the Southwest* 39: 703–857.

Búrquez, Alberto. 1998. Historical summary of the formation of the biosphere reserve El Pinacate y El Gran Desierto de Altar. In *The Sierra Pinacate*, ed. Julian D. Hayden, 74–75. Tucson: University of Arizona Press.

Búrquez, Alberto, and Carlos Castillo. 1993. *Propuesta de reserva de la biosfera El Pinacate y Gran Desierto de Altar.* Mexico City and Hermosillo, Sonora: Centro de Ecología (Universidad Nacional Autónoma de México) and Centro Ecológico de Sonora.

Camp, Martha. 1994. Fax to Carlos Nagel, Sonoran Task Force Alliance, November 15. Files of the Sonoran Institute, Tucson.

Carabias, Julia, and Bruce Babbitt. 1997. *Letter of Intent between the Department of Interior (DOI) of the United States and Secretariat of Environment, Natural Resources and Fisheries (SEMARNAP) of the United Mexican States for Joint Work in Natural Protected Areas on the United States-Mexico Border.* http://www.cerc.usgs.gov/fcc/protected_agreement.htm.

Carmony, Neil B., and David E. Brown. 1993. Epilogue: Charles Sheldon's Southwest legacy. In *The wilderness of the Southwest: Charles Sheldon's quest for desert bighorn sheep and adventures with the Havasupai and Seri Indians*, Charles Sheldon (author), 193–204. Salt Lake City: University of Utah Press.

Castillo-Sánchez, Carlos. n.d. *Highways and wildlife conservation in Mexico: The Sonoran pronghorn antelope at the El Pinacate y Gran Desierto de Altar Biosphere Reserve along the Mexico-USA border.* Hermosillo, Sonora, México: Reserva de la biosfera de El Pinacate y Gran Desierto de Altar, Instituto Nacional de Ecología. http://www.bts.gov/smart/cat/ICOWET_III/POMISULAfinal.pdf.

CCI [Comité Consenso Internacional]. 1993a. Letter to Senator Dennis DeConcini [no letterhead]. Files of the International Sonoran Desert Alliance, Ajo, AZ.

CCI [Comité Consenso Internacional]. 1993b. Letter to Senator Dennis DeConcini [Sonoran Institute letterhead]. Files of the International Sonoran Desert Alliance, Ajo, AZ.

CCI [Comité Consenso Internacional]. 1993c. *Meeting minutes for June 5 meeting.* Files of the International Sonoran Desert Alliance, Ajo, AZ.

CCR [Comité Consenso Regional]. 1993. *Meeting minutes for April 2 meeting.* Files of the International Sonoran Desert Alliance, Ajo, AZ.

CCR [Comité Consenso Regional]. c.1993a. *Cuestionario.* Files of the International Sonoran Desert Alliance, Ajo, AZ.

CCR [Comité Consenso Regional]. c.1993b. *Logo Art Contest.* Files of the International Sonoran Desert Alliance, Ajo, AZ.

CensusScope. n.d.-a. *Phoenix-Mesa, AZ, population growth.* http://www.censusscope.org/us/m6200/chart_popl.html.

CensusScope. n.d.-b. *Tucson, AZ, population growth.* http://www.censusscope.org/us/m8520/chart_popl.html.

Christensen, John R. 1994. Letter to Lorraine M. Eiler, July 8. Lower Gila Resource Area Area Manager. Bureau of Land Management, U.S. Department of the Interior. Files of the International Sonoran Desert Alliance, Ajo, AZ.

Chronicle of Community. 1997. A conversation with Luther Propst. 1, no. 3: 32–36.

Cisneros, Jose A., and Valerie J. Naylor. 1999. Uniting la frontera: The ongoing efforts to establish a transboundary park. *Environment* (April): 12–20ff.

Coate, Roger A. 1988. *Unilateralism, ideology, & U.S. foreign policy: The United States in and out of UNESCO.* Boulder, CO: L. Rienner.

Comité Consenso Sonorense Internacional. 1993. *Discussion paper: Whether to organize as a non-profit corporation.* June 5. Files of the International Sonoran Desert Alliance, Ajo, AZ.

Condo, Antoinette. 1996. 1995 Biosphere reserve managers workshop. *Park Science* (Winter): 8.

Condo, Antoinette J., and Sarah H. Bishop. 1995. *Biosphere reserves in action: Case studies of the American experience.* U.S. Department of State Publication 10241. Washington, DC: U.S. National Committee for Man and the Biosphere. http://www.state.gov/www/global/oes/cases.html.

Cornelius, Steve. 1998. *Fragmentation of natural resource management in the Sonoran Desert: Issues and resolutions.* February. Tucson, AZ: Sonoran Institute.

CSI [Consenso Sonorense Internacional]. 1993a. *Bylaws of Consenso Sonorense Internacional* [Draft]. Files of the International Sonoran Desert Alliance, Ajo, AZ.

CSI [Consenso Sonorense Internacional]. 1993b. *Draft Articles of Incorporation of Consenso Sonorense Internacional.* Files of the International Sonoran Desert Alliance, Ajo, AZ.

Curtis, James R. 1992. A closer look: The U.S.-Mexico border: A line, or a zone? In *Regional geography of the United States and Canada,* ed. Tom L. McKnight and C. Langdon White, 376–377. Englewood Cliffs, NJ: Prentice Hall.

de Klemm, Cyrille, and Clare Shine. 1993. *Biological diversity, conservation, and the law.* Gland, Switzerland: IUCN.

Eiler, Lorraine. 1992. Hia-Ced O'odham statement to the regional townhall meeting. Files of the International Sonoran Desert Alliance, Ajo, AZ.

Emanuel, Robert. 2000. *Human dimensions of the Sonoran Desert ecoregion: A summary report.* Tucson, AZ: Sonoran Institute.

Environment Committee of the Arizona-Mexico Commission, ed. 1988. *Simposio de investigación sobre la zona ecológica de El Pinacate.* Hermosillo, Sonora, Mexico: Author.

Environmental Education Exchange. n.d. *Juntos: Intercultural environmental studies in the western borderlands.* http://www.eeexchange.org/ee/HTML/WNewJuntos.htm.

Erickson, Winston P. 1994. *Sharing the desert: The Tohono O'odham in history.* Tucson: University of Arizona Press.

Ezcurra, Exequiel, Luis Bourillon, Antonio Cantu, Maria Elena Martinez, and Alejandro Robles. 2002. Ecological conservation. In *A new island biogeography of the Sea of Cortés,* ed. Ted J. Case, Martin L. Cody, and Exequiel Ezcurra. New York: Oxford University Press.

Ezcurra, Exequiel, M. Equihua, J. López Portillo, and E. Lagunas. 1982. El Pinacate: Vegetacion y medio ambiente. In *VI Simposio sobre el medio ambiente del Golfo de California.* Instituto Nacional de Investigaciones Forestales, Hermosillo, Sonora, Mexico, 68–73.

Ezcurra, Exequiel, Marisa Mazari-Hiriart, Irene Pisanty, and Adrián Guillermo Aguilar. 1999. *The basin of Mexico: Critical environmental issues and sustainability.* New York: United Nations University Press.

Felger, Richard Stephen, Bill Broyles, Michael Wilson, Gary Paul Nabhan, and Dale S. Turner. 2006. Six grand reserves, one Sonoran Desert. In *Dry borders: Great natural reserves of the Sonoran Desert,* ed. Richard Stephen Felger and Bill Broyles. Salt Lake City: University of Utah Press.

Fisher, Richard D. 1994. *National parks of northern Mexico.* Tucson, AZ: Sunracer Publications.

Fleming, Theodore H. 2004. Nectar corridors: Migration and the annual cycle of lesser long-nosed bats. In *Conserving migratory pollinators and nectar corridors in western North America,* ed. Gary Paul Nabhan, 23–42. Tucson: University of Arizona Press and Arizona-Sonora Desert Museum.

Fletcher, Susan R. 1999. *Biosphere reserves and the U.S. MAB Program.* Congressional Research Service Report for Congress. http://www.cnie.org/nle/crsreports/biodiversity/biodv-28.cfm.

García Blanco, Marco Antonio. 1995. Certificado de Constitucion de Sociedades Extranjeras. El C. Cónsul de México en Tucson, Arizona, Estados Unidos de América: Secretaría de Relaciones Exteriores, México. Files of the International Sonoran Desert Alliance, Ajo, AZ.

Gay, A. E. 1992. Letter to Windy [sic] Laird, November 18. Files of the International Sonoran Desert Alliance, Ajo, AZ.

Gilbert, Vernon C. 2004. *The US MAB concept and Program—A chronology addressing biosphere reserves.* United States Biosphere Reserves Association. http://www.rmrs.nau.edu/usamab/MAB_web_documents/US%20MAB_Biosphere%20Reserve%20Chronology.pdf.

Gómez-Pompa, Arturo, Rodolfo Dirzo, and Andrea Kaus. 1994. Letter of Accordance to U.S. Secretary of the Interior Bruce Babbitt. Files of the International Sonoran Desert Alliance, Ajo, AZ.

Gómez-Pompa, Arturo, and Andrea Kaus. 1998. From prehispanic to future conservation alternatives: Lessons from Mexico. *NAS Colloquium: Plants and Population: Is there time? December 5–6, 1998.* http://www.lsc.psu.edu/nas/Speakers/Gomez-Pompa%20manuscript.html.

Gregg, William P. Jr. 1991. MAB biosphere reserves and conservation of traditional land use systems. In *Biodiversity: Culture, conservation, and ecodevelopment,* ed. Margery L. Oldfield and Janis B. Alcorn, 274–294. Boulder, CO: Westview Press.

Gregg, William P. Jr. 1992. Letter to Wendy Laird. Files of the International Sonoran Desert Alliance, Ajo, AZ. Reference no. N16(490).

Grumbine, R. E. 1992. *Ghost bears: Exploring the biodiversity crisis.* Washington, DC: Island Press.

Guia Roji, S. A. de C. V. 1998. *Gran atlas de carreteras.* Mexico City.

Halffter, Gonzalo. 1984. Conservation, development and local participation. In *Ecology in practice,* ed. Francesco di Castri, F. W. G. Baker, and M. Hadley, 428–436. Dublin, Ireland: Tycooly International and UNESCO.

Halffter, Gonzalo. 1985. Biosphere reserves: Conservation of nature for man. *Parks* 10, no. 3: 15–18.

Halffter, Gonzalo, and Exequiel Ezcurra. 1982. El Pinacate como area de reserva. In *VI Simposio sobre el medio ambiente del Golfo de California, 8–12 April 1981*, 265–270. Hermosillo, Sonora, Mexico: Instituto Nacional de Investigaciones Forestales.

Hartmann, William K. 1989. *Desert heart: Chronicles of the Sonoran Desert*. Tucson, AZ: Fisher Books.

Hayden, Julian D. 1988. The Sierra Del Pinacate. In *Simposio de investigación sobre la zona ecológica de El Pinacate*, 52–56. Hermosillo, Sonora, Mexico: Environment Committee of the Arizona-Mexico Commission.

Hayden, Julian D. 1998. *The Sierra Pinacate*. Tucson: University of Arizona Press.

Hecla Mining Company. n.d. History: 1997-2004. http://www.hecla-mining.com/hist1997.html.

Helvarg, David. 1994. *The war against the greens: The "wise-use" movement, the new right, and anti-environmental violence*. San Francisco: Sierra Club Books.

Inter-Hemispheric Resource Center. 1997. Border Briefs: Arizona, Sonora step up binational desert protection. *BorderLines* 5, no. 4 (April). http://americaspolicy.org/borderlines/1997/bl34/bl34bb.html.

ISDA. 1993a. *Report on the International Sonoran Desert Alliance Committee (ISDA)*. Files of the International Sonoran Desert Alliance, Ajo, AZ.

ISDA. 1993b. *Summary minutes for the International Sonoran Desert Alliance meeting held in Sonoita [sic], Sonora, Mexico, October 25*. Files of the International Sonoran Desert Alliance, Ajo, AZ.

ISDA. 1994a. *Accomplishments: Calendar year 1994*. Files of the International Sonoran Desert Alliance, Ajo, AZ.

ISDA. 1994b. Bridging borders: A cross-border exchange [International Sonoran Desert Conference report]. *Sonoran Journal* (Spring).

ISDA. 1994c. Summary minutes for the International Sonoran Desert Alliance meeting held in Purto Peñasco, Sonora, Mexico. Files of the International Sonoran Desert Alliance, Ajo, AZ.

ISDA. c.1994. Conference, "Bridging Borders: A Cross-Border Exchange." Press release. Files of the International Sonoran Desert Alliance, Ajo, AZ.

ISDA. 1995a. *Draft 501(c)(3) application*. Files of the International Sonoran Desert Alliance, Ajo, AZ.

ISDA. 1995b. *Summary minutes for the International Sonoran Desert Alliance meeting held at Sonoyta, Sonora, Mexico*. Files of the International Sonoran Desert Alliance, Ajo, AZ.

ISDA. 1996. *Bylaws*. Files of the International Sonoran Desert Alliance, Ajo, AZ.

ISDA. c.1996a. *The International Sonoran Desert Alliance [with logo]*. Pamphlet. Files of the International Sonoran Desert Alliance, Ajo, AZ.

ISDA. c.1996b. *The International Sonoran Desert Alliance [with picture of O'odham baskets]*. Pamphlet. Files of the International Sonoran Desert Alliance, Ajo, AZ.

ISDA. 2000a. Colorado River water demands being addressed by user states. *Vista* [newsletter of the International Sonoran Desert Alliance] (Summer): 9.

ISDA. 2000b. Delta region focus of attention: Long neglected resource needs it. *Vista* [newsletter of the International Sonoran Desert Alliance] (Summer): 10.

ISDA. 2000c. No to "Growing Smarter" Prop. 100? but, why? *Vista* [newsletter of the International Sonoran Desert Alliance] (Summer): 5.

ISDA. n.d. *Roots Raices Ta:tk: Three languages, one meaning.* Pamphlet. Files of the International Sonoran Desert Alliance, Ajo, AZ.

ISDA Board of Directors. 1994. Letter to community members of the Sonoran Desert, September 21. Files of the Sonoran Institute, Tucson, AZ.

Joquin, Angelo Sr. 1988. The Tohono O'odham Nation: A position statement. In *Simposio de Investigación sobre la zona ecológica de El Pinacate*, 13–14. Hermosillo, Sonora, Mexico: Environment.

Keiller, Douglas. 1993. Letter to Bruce Wright, November 22. Files of the International Sonoran Desert Alliance, Ajo, AZ.

Keystone Center. 1996. *The Keystone national policy dialogue on ecosystem management: Final report.* Keystone, CO.

Keystone National Policy Dialogue on Ecosystem Management. 1995. Initiative: International Sonoran Desert Alliance (ISDA). In *Southwest Regional Meeting Background Materials*, ed. Keystone National Policy Dialogue on Ecosystem Management. Tucson, Arizona.

Kingsolver, Barbara. 1996. *Holding the line: Women in the great Arizona mine strike of 1983.* New York: ILR Press.

Krech, Shepard. 1999. *The ecological Indian: Myth and history.* New York: W.W. Norton.

Laird, Wendy. 1992. Memo to participants in the organizational meeting for the western Sonoran Desert Border Committee, draft minutes of the organizational meeting of the Regional Steering Committee, December 1. Files of the International Sonoran Desert Alliance, Ajo, AZ.

Laird, Wendy. 1993a. *Draft minutes for Comité Consenso Regional Steering Committee meeting [Sonoran Institute, February 19].* Files of the International Sonoran Desert Alliance, Ajo, AZ.

Laird, Wendy. 1993b. *Draft minutes: Second meeting of the Regional Steering Committee, January 8, 1992 [sic; actual date is 1993].* Files of the International Sonoran Desert Alliance, Ajo, AZ.

Laird, Wendy. 1993c. Letter to Louie Walters, November 28. Files of the International Sonoran Desert Alliance, Ajo, AZ.

Laird, Wendy. 1993d. Memo to Comité Consenso, April 5. Files of the International Sonoran Desert Alliance, Ajo, AZ.

Laird, Wendy. 1993e. Memo to Comité Consenso Regional, March 11. Files of the International Sonoran Desert Alliance, Ajo, AZ.

Laird, Wendy. 1993f. *Notes for 2 April Meeting [Comité Consenso Regional].* Files of the International Sonoran Desert Alliance, Ajo, AZ.

Laird, Wendy. 1993g. *Press advisory [re: June 5 Meeting of Comité Consenso Internacional].* Files of the International Sonoran Desert Alliance, Ajo, AZ.

Laird, Wendy. 1993h. *Summary minutes for the International Sonoran Desert Alliance meeting of November 12.* Files of the International Sonoran Desert Alliance, Ajo, AZ.

Laird, Wendy. 1993i. *Summary minutes for the International Sonoran Desert Alliance (formerly known as the Comité Consenso Internacional) meeting of July 30.* Files of the International Sonoran Desert Alliance, Ajo, Arizona.

Laird, Wendy. 1993j. *Summary minutes for the International Sonoran Desert Alliance (ISDA) (formerly known as the Comité Consenso Internacional) meeting of September 20.* Files of the International Sonoran Desert Alliance, Ajo, AZ.

Laird, Wendy. 1994a. Building successful partnerships in the Sonoran Desert. In *Southwest Regional Meeting background materials (1995)*. Keystone, CO: Keystone National Policy Dialogue on Ecosystem Management.

Laird, Wendy. 1994b. Building successful partnerships in the Sonoran Desert: An alliance is born. *Bajada* (April): 10–11.

Laird, Wendy. 1994c. Letter to Barry Gartell, June 21. Files of the International Sonoran Desert Alliance, Ajo, AZ.

Laird, Wendy. 1994d. Letter to Enriqueta Velarde, September 3. Files of the International Sonoran Desert Alliance, Ajo, AZ.

Laird, Wendy. 1994e. Letter to Jorge Luis Gamboa, April 28. Files of the International Sonoran Desert Alliance, AZ.

Laird, Wendy. 1994f. Partnership in the Sonoran Desert: An alliance is born. *Wild Forest Review* (September-October): 73–74.

Laird, Wendy. 1995. Letter to Robert Hardaker, January 3. Files of the International Sonoran Desert Alliance, Ajo, AZ.

Laird, Wendy. n.d. Press release [re: *Land Use Changes in the Western Sonoran Desert Border Area Conference*]. Files of the International Sonoran Desert Alliance, Ajo, AZ.

Laird, Wendy, and John Anderson. 1996a. Building the International Sonoran Desert Alliance. *Arid Lands Newsletter* (Spring-Summer). http://ag.arizona.edu/OALS/ALN/aln39/laird.html.

Laird, Wendy, and John Anderson. 1996b. One man's story. *Arid Lands Newsletter* (Spring-Summer). http://ag.arizona.edu/OALS/ALN/aln39/isdasidebar.html.

Laird, Wendy, Joaquin Murrieta-Saldivar, and John Shepard. 1997. Cooperation across borders: A brief history of biosphere reserves in the Sonoran Desert. *Journal of the Southwest* 39, no. 3–4: 307–313.

Laird, Wendy, Gary P. Nabhan, and Luther Propst. 1995. Letter to D. Dean Bibles, January 21. Files of the International Sonoran Desert Alliance, Ajo, AZ.

Laird, Wendy, and Carlos Nagel. 1993a. Memo to ISDA members, November 2. Files of the International Sonoran Desert Alliance, Ajo, AZ.

Laird, Wendy, and Carlos Nagel. 1993b. Memo to miembros de la Alianza, December 1. Files of the International Sonoran Desert Alliance, Ajo, AZ.

Laird, Wendy, and Carlos Nagel. 1993c. Memo to participants of the International Sonoran Desert Alliance, October 13. Files of the International Sonoran Desert Alliance, Ajo, AZ.

Laird, Wendy, and Carlos Nagel. 1994a. Memo to Board of Directors, August 15. Files of the International Sonoran Desert Alliance, Ajo, AZ.

Laird, Wendy, and Carlos Nagel. 1994b. Memo to Members, International Sonoran Desert Alliance, December 13. Files of the International Sonoran Desert Alliance, Ajo, AZ.

Laird, Wendy, and Carlos Nagel. 1994c. Memo to Sonoran Desert Alliance, February 28. Files of the International Sonoran Desert Alliance, Ajo, AZ.

Laird, Wendy, and Carlos Nagel. 1994d. Memo to Sonoran Desert Alliance Members [marked 1993], January 29. Files of the International Sonoran Desert Alliance, Ajo, AZ.

Lamb, Henry. 1997. Comments on the American Land Sovereignty Act. Testimony before the Committee on Resources, U.S. House of Representatives, June 10. Hollow Rock, TN: Environmental Conservation Organization. http://resourcescommittee.house.gov/archives/105cong/fullcomm/jun10.97/ lamb.htm.

La Rue, Steve. 1993. Mexico protects gulf, desert areas: Biological preserves declared by Salinas. *San Diego Union Tribune*, June 11. http://www.sci.sdsu.edu/salton/BiolPreservesbySalinas.html.

LeDuff, Charlie. 2001. A perilous 4,000-mile passage to work. *New York Times*, May 29, A1.

Lieberman, Susan. 2000. Conservation connections in a fragmented desert environment: The U.S.-Mexico border. *Natural Resources Journal* 40, no. 4: 989–1016.

Lincoln Institute of Land Policy and Sonoran Institute. 1992. *Land Use Changes in the Western Sonoran Desert Border Area: A regional forum*. Ajo, AZ: Lincoln Institute of Land Policy, Sonoran Institute, and Office of Arid Lands Studies, University of Arizona.

Luke Air Force Base. n.d. Fact sheet on Gila Bend Auxiliary Field (GBAFAF) at the Barry M. Goldwater Range (BMGR), Gila Bend, AZ. http://66.102.7.104/search?q=cache:76uHsl-yicEJ:www.luke.af.mil/ urbandevelopment/docs/Fact%2520Sheet%2520on%2520Gila%2520Bend .doc+luke+gunnery+range+goldwater&hl=en.

Manes, Christopher. 1990. *Green rage: Radical environmentalism and the unmaking of civilization*. Boston: Little, Brown.

McGuire, Randall H. 1995. Working together on the border. *Society for American Archaeology Bulletin* 13, no. 5. http://www.saa.org/publications/saabulletin/13-5/SAA8.html.

McHugh, Lois. 2000. *World Heritage Convention and U.S. National Parks*. Washington, DC: Congressional Research Service. Updated July 17. Report for Congress 96-395. http://www.ncseonline.org/NLE/CRSreports/international/inter-1.cfm?&CFID=665507&CFTOKEN=22441368.

Muro, Mark. 1997. Sonoran desert protection: New trans-border conservation network gets blessing of U.S., Mexican officials. *Planet ENN: The Magazine for the Environment*. http://www.eco.utexas.edu/~archive/chiapas95/1997.02/msg00183.html.

Murrieta-Saldivar, Joaquin. 1998. National parks in the context of biosphere reserves. In *First conference on research and management in southern Arizona park areas: Extended abstracts*, 64–67. Tucson: Organ Pipe Cactus National Monument and Cooperative Park Studies Unit.

Nabhan, Gary Paul. 1994. Letter to Task Force Member. Files of the Sonoran Institute, Tucson, AZ.

Nabhan, Gary Paul. 1995. *Completion of the Sonoran Desert Biosphere Reserve Network along the U.S./Mexico Border: Proposal for UNESCO International Biosphere Reserve Conference [Seville, Spain]*. Files of the International Sonoran Desert Alliance, Ajo, AZ.

Nabhan, Gary Paul. 1996. *Completion of the Sonoran Desert Biosphere Reserve Network along the U.S./Mexico Border: A proposal for the U.S. MAB Directorate, 1996*. Files of the International Sonoran Desert Alliance, Ajo, AZ.

Nabhan, Gary Paul. 2000. Interspecific relationships affecting endangered species recognized by O'odham and Comcáac cultures. *Ecological Applications* 10, no. 5: 1288–1295.

Nabhan, Gary Paul. 2002. *Coming home to eat: The pleasures and politics of local foods*. New York: W.W. Norton.

Nabhan, Gary Paul, Wendy Hodgson, and Frances Fellows. 1989. A meager living on lava and sand? Hia Ced O'odham food resources and habitat diversity in oral and documentary histories. *Journal of the Southwest* 31, no. 4: 508–533.

Nabhan, Gary Paul, Caroline Wilson, and David Fuller. n.d. Letter to writer/photographer. Files of the Sonoran Institute, Tucson, AZ.

Nagel, Carlos. 1988. *Report on treaties, agreements, and accords affecting natural resource management at Organ Pipe Cactus National Monument*. Special Report No. 8. Tucson: Cultural Exchange Service (Cooperative National Park Resources Studies Unit, University of Arizona).

Nagel, Carlos. 1992a. Letter to Charles Fausold, October 28. Files of the International Sonoran Desert Alliance, Ajo, AZ.

Nagel, Carlos. 1992b. *Program evaluation: Land Use Changes in the Western Sonoran Desert Border Area: A Regional Forum* [Ajo, Arizona, November 16]. Files of the International Sonoran Desert Alliance, Ajo, AZ.

Nagel, Carlos. 1994. *Minutes for the International Sonoran Desert Alliance Board of Directors meeting, March 13*. Files of the International Sonoran Desert Alliance, Ajo, AZ.

Nagel, Carlos, and Reggie Cantú. 2001. Letter to "ISDA Friend" [with attachment: "ISDA highlights"], October 17. Ajo, AZ.

Nimkin, David A. n.d. *Sustainable business development opportunities in the western borderlands*. Files of the International Sonoran Desert Alliance, Ajo, AZ.

Oppenheimer, Andres. 1996. *Bordering on chaos: Guerrillas, stockbrokers, politicians, and Mexico's road to prosperity*. Boston: Little, Brown.

Organ Pipe Cactus National Monument and Sonoran Institute. 1994. Cooperative agreement between Organ Pipe Cactus National Monument and Sonoran Institute. Files of the Sonoran Institute, Tucson, AZ.

Ortiz Garay, Andrés. 1999. Aspectos socioculturales en la gestión ambiental: El caso de minera hecla en Quitovac, Sonora. *Gaceta Ecológica*, no. 52, 53–58.

Page, Jake. 2003. *In the hands of the Great Spirit: The 20,000-year history of American Indians*. New York: Free Press.

Parfit, Michael. 1990. Earth First!ers wield a mean monkey wrench. *Smithsonian* (April): 184.

Parra-Salazar, Ivan E., Maria del Pilar Albar-Reynoso, and Juan Carlos Barrera-Guevara. 1997. *Proceedings of the 8th U.S./Mexico Border States Conference on Recreation, Parks and Wildlife*, February 26–March 1, Hermosillo, Sonora, Mexico. http://www.planeta.com/ecotravel/border/guaymas/97memindex.html.

Pearson, Gina. 1998. *Organ Pipe Cactus National Monument tri-national management challenges and opportunities for cooperation with Mexico and the Tohono O'odham Nation: A historical perspective.* Organ Pipe National Monument: U.S. Department of the Interior and Instituto Nacional de Ecología (SEMARNAP).

Phillips, Steven J., and Patricia Wentworth Comus, eds. 2000. *A natural history of the Sonoran Desert.* Tucson, AZ: Arizona–Sonora Desert Museum Press.

Planeta.com. 1995. Profile: The International Sonoran Desert Alliance. http://www.planeta.com/ecotravel/mexico/sonoran/0895isda.html.

Polzer, Charles W. 1998. *Kino, a legacy: His life, his works, his missions, his monuments.* Tucson: Jesuit Fathers of Southern Arizona.

Portman, Karla, and Wendy Laird. n.d. Press release [re: Bridging Borders: A Cross-Border Exchange Conference]. Files of the International Sonoran Desert Alliance, Ajo, AZ.

Preston, William, Edward S. Herman, Herbert I. Schiller, and Institute for Media Analysis. 1989. *Hope & folly: The United States and Unesco, 1945–1985.* Minneapolis: University of Minnesota Press.

Propst, Luther. 1994. Letter to Alastair H. Summers, Minera Hecla. Files of the International Sonoran Desert Alliance, Ajo, AZ.

Propst, Luther. 2003. *Living within limits in the West: Envisioning a West that works.* Tucson, AZ: Sonoran Institute. http://www.sonoran.org/pdfs/Rocky%20Mtn%20Land%20Inst10-16-03-transcript.pdf.

Pulliam, H. Ronald. 1994. Letter to Dr. Gary Paul Nabhan, October 24. Files of the International Sonoran Desert Alliance, Ajo, AZ.

Quintana Silver, Angel. 1994a. Memo to members of the Sonoran Desert Alliance, March 31. Files of the International Sonoran Desert Alliance, Ajo, AZ.

Quintana Silver, Angel. 1994b. Summary minutes for the International Sonoran Desert Alliance, April 16. Files of the International Sonoran Desert Alliance, Ajo, AZ.

Reyes-Castillo, Pedro. *Las reservas de la biosfera in México: Ensayo histórico sobre su promoción.* Instituto de Ecología, A.C. http://ecologia.uat.mx/biotam/v3n1/art1.html.

Rikoon, J. Sanford, and Theresa L. Goedeke. 2000. *Anti-environmentalism and citizen opposition to the Ozark Man and the Biosphere Reserve.* Lewiston, NY: Edwin Mellen Press.

Ripley, J. Douglas, Thomas H. Lillie, Stephen E. Cornelius, and Robert M. Marshall. 2000. The U.S. Department of Defense embraces biodiversity conservation through ecoregional partnerships in the Sonoran Desert. *Diversity* 15, no. 4: 3–5.

Rodriguez, Emilia, Roberto Cruz, and Alberto Gomez. 1988. Narrative/philosophy. In *Simposio de Investigación sobre la zona ecológica de El Pinacate,* 14–18. Hermosillo, Sonora, Mexico: Environment Committee of the Arizona-Mexico Commission.

Rowell, Andrew. 1996. *Green backlash.* New York: Routledge.

Schumacher, Robert W. c.1992. *Managing Cabeza Prieta National Wildlife Refuge in the 1990s.* Files of the International Sonoran Desert Alliance, Ajo, AZ.

Sheridan, Thomas E. 2000. Human ecology of the Sonoran Desert. In *A natural history of the Sonoran Desert,* ed. Steven J. Phillips and Patricia Wentworth Comus, 105–118. Tucson, AZ: Arizona–Sonora Desert Museum Press.

Sheridan, Thomas E., and Nancy J. Parezo. 1996. *Paths of life: American Indians of the Southwest and northern Mexico*. Tucson: University of Arizona Press.

SI and FPN. c.1993. *Consenso Sonorense Internacional*. (c. Spring). Sonoran Institute and Friends of ProNatura. Files of the International Sonoran Desert Alliance, Ajo, AZ.

Sierra Club Grand Canyon Chapter. 1993. Sonoran Border Conference. *Canyon Echo*. December.

Sifford, Belinda, and Charles C. Chester. Forthcoming. 2006. Bridging conservation across La Frontera: An unfinished agenda for peace parks along the U.S.-Mexico divide. In *Peace parks: Conservation and conflict resolution*, ed. Saleem Ali. Cambridge, Massachusetts: MIT Press.

Simonian, Lane. 1995. *Defending the land of the jaguar: A history of conservation in Mexico*. Austin: University of Texas Press.

Slovic, Scott. 2004. Gary Paul Nabhan: A portrait. In *Cross-pollinations: The marriage of science and poetry*, 75–93. Minneapolis: Milkweed Editions.

Smith, Harold, and Robert W. Schumacher. 1996. Letter to MAB Biosphere Reserve Committee. Files of the International Sonoran Desert Alliance, Ajo, AZ.

Soles, Roger E. 1998. Open letter to concerned citizens, February 27. U.S. MAB Program. http://www.mabnetamericas.org/sovereignty.html.

Sonoran Desert National Park Project. 2001. *A proposal: Mexico–United States Sonoran Desert Binational Peace Park*. Pamphlet prepared for the U.S.-Mexico Chamber of Commerce, Washington, DC.

Sonoran Institute. n.d. Case studies: International Sonoran Desert Alliance. http://www.sonoran.org/resources/casestudies/si_case_isda.html.

Sonoran Institute. 1992. Land use changes in the western Sonoran Desert region: A forum [summary of results from work group sessions]. Files of the International Sonoran Desert Alliance, Ajo, AZ.

Sonoran Institute. 1999. Tohono O'odham Nation. http://www.sonoran.org/resources/terms/si_glossary_tohono.html.

Sonoran Institute and International Sonoran Desert Alliance. n.d. Cooperative resource management [proposal to the Ford Foundation]. Files of the International Sonoran Desert Alliance, Ajo, AZ.

Steffens, Ron. 1994. Learning to speak Sonoran. *Tucson Weekly*, February 9–16.

Steinhart, Peter, and Tupper Ansel Blake. 1994. *Two eagles/dos aguilas: The natural world of the United States–Mexico borderlands*. Berkeley: University of California Press.

Stern, Kenneth S. 1996. *A force upon the plain: The American militia movement and the politics of hate*. New York: Simon & Schuster.

Tangley, Laura. 1988. A new era for biosphere reserves. *BioScience* 38, no. 3: 148–155.

Tohono O'odham Community College. n.d. *About TON*. http://www.tocc.cc.az.us/about_ton.htm.

TPECR. 1994. Minutes to 1 February meeting. Taskforce on the Protection of Ecological and Cultural Resources in the Western Sonoran Desert Borderlands. Files of the Sonoran Institute, Tucson, AZ.

TPSDBR. 1994. Update on progress as of November 1994 [attachment to letter from Caroline Wilson to Taskforce members, dated November 9]. Taskforce on proposed Sonoran Desert Biosphere Reserve. Files of the Sonoran Institute, Tucson, AZ.

Transboundary Sonora Desert Letter. 1997. Toward completion of a transboundary Sonoran Desert network of protected areas, with multi-cultural community involvement. Hermosillo, Mexico. Files of the International Sonoran Desert Alliance, Ajo, AZ.

Turner, Tom. 2002. Justice on earth: Earthjustice and the people it has served. White River Junction, VT: Chelsea Green.

Udall, Morris K. 1966. A national park for the Sonoran Desert. *Audubon Magazine*, 105–109. http://dizzy.library.arizona.edu/branches/spc/udall/sonoran_htm.html.

Udall, Stewart. 1997. Stewart Udall: Sonoran Desert National Park. Interview by Jack Loeffler. *Journal of the Southwest* 39, no. 3–4: 315–320.

UNA-USA. 1993. *Schooling for democracy: American panel on UNESCO: Reinventing UNESCO for the post–Cold War world.* New York: United Nations Association of the United States of America. http://www.unausa.org/atf/cf/%7B49C555AC-20C8-4B43-8483-A2D4C1808E4E%7D/schooling.pdf.

UNESCO. 1974. *International Co-ordinating Council of the Programme on Man and the Biosphere (MAB), Third session, Final report.* Washington, DC, September 17–29. MAB report series No. 27.

UNESCO. 2005. UNESCO-MAB biosphere reserve directory: Latin America and the Caribbean. Man and the Biosphere (MAB) Programme. http://www.unesco.org/mab/brlistlatin.htm.

UNESCO. 2000. World network of biosphere reserves. Man and the Biosphere Programme. http://www.unesco.org/mab/brlist.htm.

United Nations Environment Programme and World Conservation Monitoring Centre. 2005. World database on protected areas: Mexico. http://sea.unep-wcmc.org/wdbpa/.

U.S. Congress. n.d. *American Land Sovereignty Protection Act.* Committee on Resources, House Briefing Paper (H.R. 3752). http://resourcescommittee.house.gov/archives/104cong/fullcomm/hr3752/3752brf.htm.

U.S.-Mexico Border Field Coordinating Committee. 2001. U.S.-Mexico Sister Areas Issue Team. U.S. Department of the Interior. http://www.cerc.cr.usgs.gov/FCC/issue%20teams/US_Mexico_Sister_Areas.html.

U.S. Department of State. 1997. Governors Symington and Beltrones endorse biosphere reserve agreement. *U.S. MAB Bulletin.* http://www.state.gov/www/global/oes/bul_3_97.html.

U.S. Department of State. 1998. U.S. MAB partnerships fact sheet. Bureau of Oceans and International Environmental and Scientific Affairs. http://www.state.gov/www/global/oes/fs_partnerships.html.

U.S. Fish and Wildlife Service. n.d. *Cabeza Prieta National Wildlife Refuge* [pamphlet]. Washington, DC: U.S. Department of the Interior.

U.S. MAB Program. n.d. The U.S. Man and the Biosphere Program http://www.mabnetamericas.org/general_information/geninfo.html.

U.S. National Park Service. 1965. *Sonoran Desert National Park, Arizona: A proposal.* Southwest Region, U.S. Department of the Interior.

U.S. National Park Service. 1998. Man and the Biosphere Program and international biosphere reserve designations. Department of the Interior. http://www.nature.nps.gov/partner/mabpage.htm.

Vesilind, Priit J. 1994. The Sonoran Desert. *National Geographic* (September): 36–63.

Vidal, Rosa Maria, Lucia Grenna, and Daniel Calabrese. 2004. Strategic communication planning for a national system of protected areas, Mexico. In *Communicating protected areas (a compilation of papers on education and communication presented to the Vth IUCN World Parks Congress, Durban, South Africa, 8–17 September 2003),* ed. Denise Hamú, Elisabeth Auchinclos, and Wendy Goldstein. Gland, Switzerland: Commission on Education and Communication, World Conservation Union (IUCN).

Vista. 1999. Vol. 4, no. 1, International Sonoran Desert Alliance, Ajo, AZ.

Waldman, Carl. 1985. *Atlas of the North American Indian.* New York: Facts on File.

Walker, Marisa Paula, and Vera Pavlakovich-Kochi. 2003. *The state of the Arizona-Sonora border region: Shared pollution, shared solutions.* Southwest Consortium for Environmental Research and Policy (Border Institute V Conference). http://www.scerp.org/bi/BIV/AZSon.pdf

Walker, Steve. n.d. *El Pinacate Biosphere Reserve.* San Antonio, Texas: Nature Conservancy. http://parksinperil.org/files/page_4_el_pinacate_biosphere_reserve.pdf. Also available in ring-binder report: The Mexico program: A guide to places and projects, 2005 edition.

Wikipedia. 2004. Phoenix, Arizona. http://en.wikipedia.org/wiki/Phoenix,_Arizona.

Williams, Florence. 1994. On the borderline. *High Country News*, March 21, 1ff. http://www.hcn.org/servlets/hcn.Article?article_id=162.

Yetman, David. 1996. *Sonora: An intimate geography.* Albuquerque: University of New Mexico Press.

Zakin, Susan. 1993. *Coyotes and town dogs: EarthFirst! and the environmental movement.* New York: Viking.

Zakin, Susan. 1997. Mexico launches a green offensive. *High Country News*, March 31, 2.

Zeller, Tom. 2001. Migrants take their chances on a harsh path of hope. *New York Times*, March 18, 14.

4

Landscape Vision and the Yellowstone to Yukon Conservation Initiative

Be it through far-flung e-mails or face-to-face conversations, a fundamental goal of the Yellowstone to Yukon Conservation Initiative has been to achieve common understanding and purpose between Canadians and Americans. Yet as with other initiatives between the two countries, ignorance has never been mutual. Fortunately, Canadians tend to fall back on humor to counter their frustrations with the dearth of knowledge exhibited by their southerly neighbors.

Take, for example, the earnest Rick Mercer wandering among the ivy-laden buildings of Princeton University informing students and professors alike of the decision by Toronto's mayor to reinstate the city's polar bear hunt. Gladly signing Mercer's protest petition, the outraged Princetonians would no doubt have been mortified to learn of Mercer's identity as a popular Canadian satirist of pandemic U.S. ignorance and stereotyping of anything Canadian—including urban polar bears—and whose creative derision has endeared him to millions of Canadians. In targeting the inward-looking and economically dominant steamroller that constitutes Canada's only bordering neighbor, Mercer's shtick would come off as banal were it not so funny. For although he has come up with a creative approach, Mercer is hardly original in pointing out that Canada barely registers on the cultural and political radar screens of Americans—even many highly educated Americans strolling the yards of Princeton.

Along similar lines, many Canadians have been somewhat taken aback by the multi-year, chest-thumpingly patriotic celebrations in the United States over the two-hundredth anniversary of Lewis and Clark's first voyage across the continent north of the Mexico-U.S. border. *First?* "Hold on!" Canadians exclaimed. "We got there first" (Newman 2003). Indeed, in 1793—a full decade before Lewis and Clark set out—explorer Alexander Mackenzie set out from his northwestern outpost of Fort Chipewyan, crossed the Rocky Mountains north of what would become Alberta's Jasper National Park, and finally reached the Pacific Ocean near present-day Bella Coola, British Columbia (Hayes 2001; Mackenzie 1801). Despite the fact that Mackenzie was born and died in Great Britain—not to mention that Canada did not exist as an independent country in 1793—Canadians today claim Mackenzie as their own.

Although Mackenzie's crossing spurred Jefferson to send forth Lewis and Clark (Ronda 1989, 149–150), historians have conducted little by way of formal comparative analysis between the two explorations. Superficially, it is probably fair to say that whereas the primary purpose of both expeditions was to find a way to the Pacific coast and back, Lewis and Clark were charged with the additional task of conducting a formal evaluation of the territory they crossed. This would at least partly account for the far more voluminous extent of the journals of Lewis and Clark, which in numerous edited forms have become perhaps the seminal document in the history of the American West. For historians, politicians, the media, and the general public in the United States, Lewis and Clark's voyage remains the key reference point in the nation's expansion across the continent.

Natural historians in the United States have also relied on Lewis and Clark's journals as a *biological* reference point. This does not appear to be the case for Canadian naturalists in regard to Mackenzie's journals, and one can speculate on why this is so: Is it because Mackenzie did not so thoroughly examine the region's flora and fauna as did Lewis and Clark? Is it because Mackenzie is not the historical touchstone in Canada that Lewis and Clark are in the United States? Or is it that Mackenzie simply did not encounter as much of the "natural world" as Lewis and Clark? Unlikely as this last hypothesis might sound, it does appear to be the case with at least one part of the natural world: the bear.

As recorded in his journals, Mackenzie had relatively few encounters with bears. In his travels from just east of the Rockies to his return, he recorded a total of four sightings of bears, one of which he (or someone in his party) killed (Mackenzie 1801, 229, 230, 407). It is impossible now to identify which species of bear he came across—although he does describe two of them as "grisly and hideous"—and his journals do not indicate that he was inordinately concerned over the prospect of a bear attack.

In contrast, Lewis and Clark's Corps of Discovery had far more direct contact with bears—particularly as they approached the Rocky Mountains. In the two weeks before May 26, 1805, individual crew members escaped from two dangerous encounters with bears that they alternately described as white, brown, gray, yellow, and "grisley"—an indecisiveness that reflected the vast gradation in grizzly bear coloring as well as the corps' apparently mixed reaction to what they perceived as both a "monster" and a "gentleman" (Lewis and Clark 1904, 1:373, 2:23–34). Wrote Lewis respectfully of what would be designated a decade later as the grizzly bear: "The wonderful power of life which these animals possess renders them dreadful" (Botkin 1995, 60). But even as the bears alternately awed and terrified the Corps of Discovery, the dangerous creature was only one of an incessant flow of unfamiliar challenges. The threat of bears may have lurked only in the back of Lewis's mind when, on May 26, he scrambled up the bluffs of the Missouri River—in what eighty-four years later would become the state of Montana—and "beheld the Rocky Mountains for the first time" (Lewis and Clark 1904, 2: 78–79).

Taking the measure of the land spread before him, Lewis viewed the Rockies with relief, since their approach meant the corps was finally nearing the headwaters of the Missouri, but also with trepidation, because they had to cross those snowcapped mountains. For all the Rockies may have symbolized to Lewis, they remained largely a desperate challenge of perseverance and survival for the corps. Today, the Rocky Mountains present a much different type of challenge, one that probably would have made little sense to Lewis and Clark—or to Mackenzie, or any of their respective crews. That challenge is to protect and defend the grizzly bear, as well as other wildlife and their wilderness habitats, from extirpation and degradation. Since the late nineteenth century, those taking on this challenge have been a diverse and often disparate group of private citizens and governmental officials. Their efforts constitute an important strand of the global conservation movement, particularly as some of the earliest battles over wildlife and wilderness protection were fought in the Rocky Mountains.

From these deep roots has arisen a conservation vision encapsulated in the phrase "Yellowstone to Yukon"—or Y2Y for short. Y2Y's character is multifarious: It is a region of both biogeographic and cultural similarities. It is a network of more than three hundred conservation groups, an organized movement to protect the land, its character, and its wildlife. It is a transborder conservation organization located in Canmore, Alberta. It is a meta-icon, composed of the geographical bedrock of U.S. conservationist philosophy and the proving ground of Canadian grit and national identity. And most important, it is a broad reconceptualization of how to protect a relatively unaltered landscape. As such, the "Y2Y vision" cuts not only across the grain of each country's version of manifest destiny, but also across the relatively small-scaled grain of applied conservation on the North American continent (Gluek 1965; Soulé and Noss 1998).

The Yellowstone to Yukon Region

The Y2Y region spreads across 1.2 million square kilometers (4,680,000 square miles), an area larger than California and Texas combined. A straight line down the center of the Y2Y region runs roughly 3,200 kilometers (1,984 miles) from the Peel River (north of the Mackenzie Mountains of the Yukon and Northwest Territories) to the southern end of the Wind River range in western Wyoming (Figure 4.1). In terms of nations and subnational units, the Y2Y region encompasses large portions of the Canadian Northwest and Yukon Territories, the Canadian provinces of British Columbia and Alberta, and the U.S. states of Montana, Idaho, and Wyoming (with smaller portions extending into Oregon and Washington).

As it was originally conceived, the "Yellowstone to Yukon corridor" was not so extensive (Locke 1997, 118). Unable to find a single map that covered the entire binational region, Canadian conservationist Harvey Locke commissioned a map of the region for the first Y2Y meeting in 1993 (Boyd 1993, 2; Locke 2002). The map showed the Y2Y region as extending from the Liard River on the north to the southern border of Yellowstone National Park (YNP); the Rocky Mountain Trench formed the western edge, and the eastern border was undefined (Boyd 1993, 2). While Locke (2002) admitted to being "a bit nervous that the Y2Y idea might be seen as indefensibly big," participants at the first meeting ended up expanding the Y2Y region to cover much more of the Yukon, the mountain ranges south of YNP, most of northern Idaho, the edges of Washington and Oregon, and all of the Columbia Mountains in British Columbia (Boyd 1993, 4; Locke 2002). Notably, participants at Y2Y meetings were still actively discussing the appropriate extent of the region up until at least 1999, and its geographical extent continues to vary in detail according to who is describing it. But even as the Y2Y region has expanded and defied precision, there is little to dispute Locke's assertion that the "rough boundaries" set in 1993 "have proven pretty robust" (Locke 2002).

Fig. 4.1
Yellowstone to Yukon Ecoregion

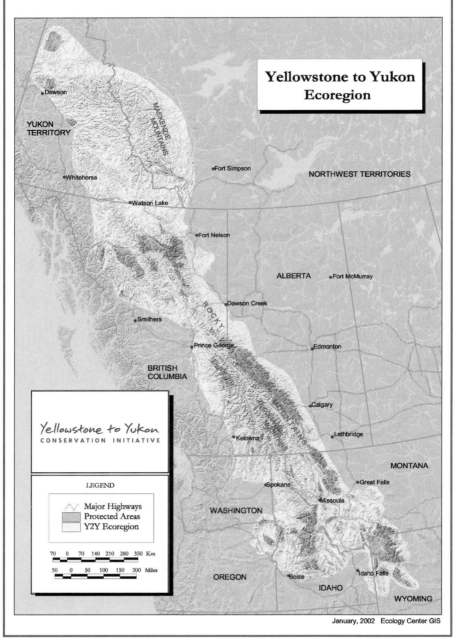

A minor point of debate among Y2Y participants has been whether the region constitutes a "coherent ecological unit." Both scientists and conservationists working under Y2Y generally have recognized that ecosystem and ecoregional boundaries are not hard and fast but rather overlap and interweave with each other. Yet there was some cause to consider the region as biologically distinct, for in his landmark treatise *The Ecology of North America*, ecologist Victor Shelford (1963, 152) classified "the Northern Rocky Mountains" as a biotic community crossing the international border—although with an outline less encompassing than that of the Y2Y region. Despite Shelford's eminent status, however, few people adopted that designation, partly as a result of competing biological, geographical, and geological labels, and perhaps partly due to the very different connotations that "the Northern Rockies" carries in each country. In the United States, the phrase generally refers to the portion of the Rocky Mountains situated in Montana, Idaho, and Wyoming (McCoy 1998; Rockwell 1998; U.S. House 1994). For Canadians, the term refers to the relatively untouched and rugged terrain in northern British Columbia (Gadd 1995). Nonetheless, a growing number of scientists and conservationists have informally come to rely on "the Northern Rockies" as synonymous to "the Y2Y region," and it is a practice that I follow in this chapter. But it is only fair to emphasize that the term is informal and, depending upon the audience, potentially confusing.

Within this landscape, no known species has gone extinct since the explorations of Mackenzie and Lewis and Clark (Y2Y Conservation Initiative n.d.-a). The region contains the "best remaining mountain wildlife habitat on the continent," and its physiography "provides habitat features that make the region exceptionally valuable to wildlife: large, ecologically intact expanses of land where natural vegetation predominates, and where ecological processes continue to operate much as they always have" (Gadd 1998, 14). The Y2Y region is particularly important for large predators whose numbers have been "drastically reduced" elsewhere (Willcox 1998, 1). Several of these are listed under the Endangered Species Act (ESA), including the grizzly bear (*Ursus arctos horribilis*), the gray wolf (*Canis lupus*), the lynx (*Lynx canadensis*), woodland caribou (*Rangifer tarandus caribou*), and the bald eagle (*Haliaeetus leucocephalus*). Several conservation groups (including participants in Y2Y) have repeatedly and unsuccessfully petitioned the U.S. Fish and Wildlife service to list the wolverine (*Gulo gulo*). The region is also well known for its relatively abundant populations of moose (*Alces alces*), elk (*Cervus elaphus*), and mule deer (*Odocoileus hemionus*), as well as pronghorn antelope (*Antilocapra americana*), and black bear (*Ursus americanus*). Each year thousands of golden eagles (*Aquila chrysaetos*) use the Northern Rockies as a migratory conduit or "eagle highway" (Tabor 1996, 2, 6; Sherrington 2005).

The Y2Y region is home to the headwaters of seven major rivers that drain into the Arctic, Pacific, and Atlantic Oceans (Y2Y Conservation Initiative n.d.-a). These river systems and other aquatic habitats are home to at least 118 fish species (Mayhood et al. 1998), several of them economically important and others endangered (Willcox 1998, 1). Threats to fish species, most notably overfishing, habitat loss and fragmentation, hybridization, and competition with introduced species, have led to a number of species being put on the endangered species list, including the bull trout (*Salvelinus confluentus*), the chinook salmon (*Oncorhynchus tshawytscha*), and the Kootenai River population of the white sturgeon (*Acipenser transmontanus*). Y2Y-affiliated organizations have lobbied to add the Westslope cutthroat trout (*Oncorhynchus clarki lewisi*) and the Arctic grayling (*Thymallus arcticus*) to the U.S. Endangered Species List (Mayhood et al. 1998).

Threats to terrestrial biodiversity in the Y2Y region include suburban and rural sprawl; logging operations and clear-cutting; direct persecution of predators; road building; mining for coal, metals, and other valuable minerals; oil and gas development; introduction of invasive species; damming and diverting rivers; grazing; fire suppression; and recreation, including off-road vehicle (ORV) use and resort development. The collective severity of these threats to biodiversity varies progressively on a north-to-south basis. Conservationist Louisa Willcox told me that when Y2Y participants started mapping the region, the geophysical contrast from north to south became clear. Whereas the northern, Canadian area consisted of wild areas punctuated with islands of development, the southern portion—mostly in the United States—was an "inverse template" of islands of wildlands in the midst of a "sea of development." Consequently, "Y2Y is largely a dream of restoration" on the U.S. side of the border. "Y2Y is about using the precautionary principle in saving connectivity in the landscape" (Tabor 1996, 28).

Borders braid and quarter the Y2Y region. Even at the time of the first European incursion into the western half of the North American continent, the southern edge of the Y2Y region formed a border—albeit a permeable one—between the plains and the plateau cultures of indigenous North America (Clark 1966). And within the Y2Y region can be found the traditional territories of thirty-one First Nation or Native American groups (Reeves 1998, 36). Onto this original palette of indigenous cultures has been spun an intricate web of borders imposed by successive waves of European immigrants. The most obvious of these is the U.S.-Canada border, which runs east-west along the 49th parallel. Next in terms of political hierarchy, the provincial border between Alberta–British Columbia extends northward along the spine of the Northern Rockies. In comparison to their U.S. state counterparts, Canadian provinces rule with much greater independence from their federal government. This greater autonomy is manifest in the provinces' ownership of the vast majority of Canada's public lands (Boyd 1993), resulting not only in a smaller role for the federal government in public land management, but also in a significant divergence in land management practices among provinces. As conservationist Gary Tabor (1996, 8–9) noted regarding the tensions between conservative Alberta and the relatively more liberal British Columbia, "Y2Y is as much about Alberta and British Columbia cooperation as it is about Canadian and American conservation cooperation." In the United States, the lower degree of individual state autonomy—combined with a strong federal presence in the U.S. Rocky Mountains—has resulted in state borders playing a relatively less visible role in the region's conservation debates.

Within and among the provinces and states, myriad other borders divvy up land between private owners, First Nation and Native American lands, and a number of government land agencies. In Canada, land management agencies include Parks Canada, Alberta Environment, British Columbia Parks, and the British Columbia Ministry of Forests. The United States has the National Park Service (NPS), the Forest Service (USFS), the Fish and Wildlife Service (FWS), the Bureau of Land Management (BLM), and the various relevant state fish, game, wildlife, and park agencies.

Y2Y's Web of Conception

Efforts to protect wildlife in the Northern Rockies date to the last quarter of the nineteenth century and have resulted in protected areas, hunting laws, and the restoration of endangered species, including "the first successful effort to save a jeopardized species—the bison" (Wuerthner 2001a, 14). Of these, the most visible has been the establishment of protected areas, mostly consisting of wildlife refuges, wilderness areas, forest reserves, and parks. The region includes two famous national parks, Yellowstone in the United States (the world's first national park, established in 1872) and Banff in Canada (established in 1885). Nine other national parks serve as "core areas" for biodiversity in the Y2Y region: Grand Teton and Glacier National Parks in the United States, and Waterton Lakes, Kootenay, Yoho, Jasper, Glacier, Mount Revelstoke, and Nahanni in Canada. The region is also home to the world's first international peace park (Waterton-Glacier established in 1932; see Chapter 2) and to the 1891 Yellowstone Park Timber Land Reserve (now considered to be the earliest predecessor of the U.S. Forest System) and the Bob Marshall Wilderness Area (composed of some of the earliest "primitive areas" established by the USFS) (Hendee and Dawson 2002, 102–103). Today, six of the largest U.S. wilderness areas outside of Alaska can be found in the region (Haines 1977, 95; Reiger 1997, 42–44). In 1992, just over 10 percent of the Y2Y region was estimated to lie under some form of protected area status, such as wildlife refuge, wilderness area, or state or provincial park (Merrill and Mattson 1998, 27). These lands have not only provided "pleasuring grounds" for millions of citizens, but are also a source of national pride in both countries. Given the prominent role of this shared mountain range in the conservation histories of both countries, "it's not surprising that the Rockies would be one of the first areas in the country where a bold new vision for large-scale conservation would be born" (Wuerthner 2001a, 14).

"Yellowstone to Yukon" stands on top of—indeed, because of—these deep historical foundations. No doubt, the historical cachet of the names "Yellowstone" and "Yukon" has helped to propel Y2Y into the conservation limelight. Yet even as Y2Y descends from more than a century's worth of conservation efforts (a heritage we will further explore later on), its direct origins can be traced to the interweaving of several scientific and political threads during the 1980s (Johns 1998). While each will be examined in turn, one particular thread stands out in the form of one individual most often cited as being the principal weaver of Y2Y: Canadian attorney Harvey Locke (Lowey 1997; see also, for example, Mitchell 1998; Newquist 2000). Although there is some disagreement, the preponderance of the evidence points to Locke as the first person to consciously juxtapose the words "Yellowstone" and "Yukon."

Locke brought to his passion for the Canadian Rockies a deep understanding of the region's politics. After a 1989 run for a seat in the Alberta

Legislative Assembly (where he did "well but didn't win" in the predominantly conservative province), he became president of the Alberta Liberal Party from 1996 to 1998. In the early 1990s, he was president of the Canadian Parks and Wilderness Society (CPAWS), one of the more influential conservation organizations in Canada. The combination of skills and interests was potent: "For nearly 20 years," one colleague wrote, "Harvey Locke's name has been synonymous with the battle to protect parks and wilderness in western Canada" (Legault 1999).

In his day job as a trial lawyer for a Calgary law firm, Locke was paying close attention to research on large carnivores in the Northern Rockies. Mulling over the concept of "wildlife connectivity," he later recounted to me that he had been:

> sitting in my basement thinking about these issues and had started scratching a map of big landscapes you might preserve in North America—just based on my travel and my own knowledge. And I'd drawn a big line around the Rocky Mountains, around the Great Lakes–Saint Lawrence forests, and a big line around the Great Basin Desert and a big line on the Adirondack-Appalachian chain because I'd been in the Smoky Mountains recently. Then I kind of just put it aside.

Then in 1993, he went on two backcountry trips, one to the Willmore Wilderness Park north of Jasper National Park, the other to the Muskwa-Kechika, a vast and relatively undeveloped portion of northern British Columbia that would become a focal area for Y2Y (see discussion below). On those trips it struck him that "a lot of people thought then, and probably still do, that the Rockies kind of ended in Jasper." Being out in these areas north of Jasper National Park for the first time expanded his conceptualization of a cohesive mountainous region stretching all the way to the Yukon Territory. But it was only the second trip, led by outfitter and conservationist Wayne Sawchuk, that Locke says he first put the words "Yellowstone to Yukon" on paper and thought it was a "real idea." In response to my pressing him on exactly how he had come to put these words together, he responded, "Well, if you're looking for the moment of crystallization, my best memory is sitting by the campfire in the Northern Rockies with Wayne Sawchuk. And I started writing on the edge of a map an article about how this stuff would all hang together. . . . And I wrote the words 'Yellowstone to Yukon.'" Sawchuk corroborated Locke's account, recalling that he first heard the phrase from Locke at the end of a long campfire conversation in which they had been musing over how to protect large areas.

Based on the text he had written on the map in the backcountry, Locke drafted a "discussion paper" for what is now commonly described as the first Y2Y meeting, held in 1993 in Kananaskis Country south of Banff National Park (Greater Ecosystem Alliance 1994; Locke 2002). David Johns (1998) described that draft as the "Locke vision statement," while George Wuerthner (2001a, 14) called it "a somewhat more modest proposal" —"more modest" referring to the relatively limited geographic extent that Locke proposed at the 1993 meeting. Locke then polished the draft for publication in CPAWS's magazine, *Borealis*, in the spring of 1994 (Locke 1994). The essential argument of the essay is that the Y2Y region constitutes a cohesive entity that requires and deserves protection. "Nowhere else in North America will you find significant numbers of people living adjacent to an area that has retained its large carnivores," Locke pointed out—quickly adding the important proviso that "these high country ecosystems are not managed in a way that recognizes their seminal importance to life in North America" (Locke 1994, 19). He concluded that protecting the biodiversity of the Northern Rockies would require far more than the current system of relatively small and isolated protected areas. It was a message that he would ceaselessly reiterate in publications and presentations over the next decade.

A few Y2Y participants and observers have expressed skepticism of Locke's account, with the most strident asserting that "the mythology that has grown up about Harvey sitting around a campfire and suddenly the idea struck him—it's just part of this dog-and-pony show about the Y2Y." That particular skeptic was likely aware of a more famous campfire conversation that reportedly led to the creation of Yellowstone National Park, a story that turns out to have little historical foundation (Schullery and Whittlesey 2003). In contrast, however, Locke's "campfire story" is well corroborated, and there is no indication that the map on which he scrawled "Yellowstone to Yukon" (which I have examined) is a fabrication.

Yet Locke himself has come to downplay his own campfire story. He is no doubt aware that most Y2Y participants would agree with economist Ray Rasker of the Sonoran Institute that whoever "first uttered the words Y2Y is immaterial." He is thus careful in public forums to ascribe the origins of Y2Y to "many Canadian and American conservationists and biologists" (Locke 1997, 119) or simply to state that Y2Y was "based on conservation biology and a love of wild place" (Locke 1998, 255). In conversation, he more elaborately describes his wilderness epiphany as nothing more than a synthesis of other people's work:

> It's a creation myth, so it's useful in that sense. But really what happened is that I labeled what people were thinking anyway. All of the threads were out there, and the emotional relationship to this landscape above all others in the context of wilderness and wildlife was out there. What I did was marshal, and I was probably well equipped to do that because of my years of doing these complicated environmental cases involving multiple expert witnesses—from plant experts to water experts to flower experts to bear experts to health experts—and making that into a coherent mosaic that could then persuade a decision-making panel of engineers and stuff that "x" was the right result. So as the synthesizer and articulator of these yearnings and ideas and intellectual concepts, I was important. But I did not invent any of the things that were synthesized; I packaged them.

Emphasizing the point of having just "labeled what people were thinking anyway," Locke pointed to a particular passage in an article by the noted author Wallace Stegner. In examining the historical origins of the American conservation movement, Stegner (1990) asserted that "the tracing of ideas is a guessing game. We can't tell who first had an idea; we can only tell who first had it influentially, who formulated it in a striking way and left it in some form, poem or equation or picture, that others could stumble upon with the shock of recognition." Yet, as Stegner would have agreed, tracing the roots of an idea is sometimes a guessing game worth playing—particularly when, as in the case of Y2Y, there is a substantial intellectual heritage to be explored.

To use Locke's own words, what exactly was he "packaging" and "labeling"? Both in the early *Borealis* article and in person, he articulates three principal influences that led to Y2Y. First, and most important, he points to the broad application of the science of conservation biology to the biodiversity of the Northern Rockies. As noted above, he had absorbed the principal tenets of that burgeoning science while pressing several lawsuits in which he called on several well-known wildlife biologists as expert witnesses. In running those cases, Locke said, "I had to get my brain into that subject matter pretty deeply. . . . I learned a lot about what the basic ecological questions are and this whole question of movement and connectivity—and from a really applied way. So I became interested in the intersection between wildlife and wilderness and the science foundations connecting to the sort of more advocacy for land-use stuff."

Emanating out of the "new" science of conservation biology were several fundamental principles that Locke implicitly and explicitly referred to in his writings and presentations on Y2Y. The *principle of connectivity* held that interconnecting wildlife corridors are necessary to maintain large carnivores in isolated protected areas. The *principle of ecological processes* called for the maintenance of predator-prey and mutualistic relationships, genetic differentiation between populations, hydrological and nutrient cycles, and other large-scale periodic events and cycles. And the *principle of umbrella species* held that the protection of sufficient habitat for large carnivores should in turn protect the majority of other species.

As a second influence, Locke cites a number of specific research efforts in the Northern Rockies that led to the conception of Y2Y. Several empirical studies on wolves, for example, demonstrated that the animals traversed wide swaths of land—an indication that the "Canadian Rockies are part of one gigantic linear ecosystem" (Locke 1994, 19). Those research efforts were led by eminent carnivore biologists Diane Boyd and Paul Paquet, both of whom became involved in Y2Y early on. In 1992, Paquet had radio-collared a female wolf he named Pluie, the French word for rain since it was raining when he found her (Heuer 2000, 70). Over one eighteen-month period, Pluie roamed 840 kilometers (521 miles) along the Rocky Mountains, across thirty different political jurisdictions (Tabor 1996, 7). By the time she was shot by a hunter in 1995, she had traveled an area of more than 100,000 square kilometers (39,000 square miles)—more than twice the size of Switzerland (Heuer 2000, 70; Newquist 2000). References to Pluie became a common refrain in literature and presentations on Y2Y, and although Locke did not use her name in the *Borealis* article, he was most likely referring to her there.

Locke had also read a more policy-oriented study entitled *A Conservation Strategy for Large Carnivores in Canada* that was released in November 1990 by World Wildlife Fund–Canada (repackaged and released in the United States as the book *Wild Hunters* the following year) (Hummel et al. 1991). As part of the "large carnivore strategy" promoted in this report, the authors proposed the establishment of Carnivore Conservation Areas

(CCAs). Defined as "areas of sufficient size and managed in such a way to ensure long-term survival for free-ranging, minimum viable populations of large carnivores," CCAs were to be linked through "natural corridors" (Hummel et al. 1991, 176). Three out of the five proposed CCAs were located within the Y2Y region (Hummel et al. 1991, 178).

The third influence in Locke's conception of Y2Y consisted of numerous advocacy initiatives by various regional conservation organizations. As the field of conservation biology solidified during the late 1980s and early 1990s, many conservationists began to utilize the aforementioned scientific principles both to develop a more encompassing land management philosophy and to justify the protection of extensive areas. That so many people in various NGOs were already thinking along the lines of large-landscape conservation, Locke believed, was why Y2Y eventually became so widely embraced. These conservation organizations included CPAWS, the Wilderness Society, and American Wildlands, the latter of which had during that time period begun its Corridors of Life project focusing on the entire U.S. portion of the Northern Rockies (American Wildlands n.d.).

Within the world of nonprofit conservation organizations, the culmination of this approach was the Wildlands Project (TWP). Originating in conservation planning exercises first drawn up for Florida and Central America (Johns 2003b), TWP was established in 1991 by conservationists and biologists who believed that the "existing methods of preserving nature in isolated 'islands' of protection are insufficient to insure the long-term survival of North America's full complement of native plants and animals" (Wildlands Project 1999). TWP subsequently became the nexus for various "large-landscape" conservation initiatives across the North American continent (Foreman 2004).

Locke also recalled that before he put together Y2Y he had been given a special issue of the magazine *Wild Earth* that examined TWP and its "North American wilderness recovery strategy." This widely distributed special issue contained articles by several well-known conservationists, including David Foreman (1992), Michael Soulé (1992), Gary Snyder (1992), and Reed Noss (1992), as well as articles on maintaining extensive wildlands in the Adirondacks (Medeiros 1992), the southern Appalachian Mountains (Newman et al. 1992), and Central America's Mesoamerican Biological Corridor (formerly *Paseo Pantera*) (Marynowski 1992). Notably, the issue also contained an article by conservationist Mike Bader (1992), a founder of the Alliance for the Wild Rockies, on the Northern Rockies Ecosystem Protection Act (described more extensively below).

Locke found that the ideas in the special issue converged with his own, and that TWP "was just exactly congruent with the stuff that I'd been getting from my experience with these various experts." A professional colleague, Monte Hummel of WWF-Canada, turned out to be one of the founders of TWP, and Hummel introduced Locke to the other founders. It was propitious timing, since the TWP founders were scheduling a series of "vision mapping" workshops that would be held around North America between 1993 and 1995 (Boyd 1993, 1). In his position as president of the board of CPAWS, Locke worked with David Johns, then executive director of TWP, to set up one of those sessions for the Rocky Mountain region that crossed the Canada–U.S. border. Held at the Kananaskis Field Station in Kananaskis Country December 3–4, 1993, the workshop opened with the explicit recognition that this was a "historic, exciting meeting" (Boyd 1993, 1). Although Locke's Y2Y draft predates this meeting, some have stated that Y2Y came out of the workshop (e.g., Greater Ecosystem Alliance 1994). To the degree that Y2Y is an initiative backed by a group of individuals and organizations, the 1993 Kananaskis meeting might reasonably be described as Y2Y's institutional origin—particularly in as much as it was here that people began to think through the full geographic extent of Y2Y, reaffirming "the need to think big and long term" (Johns 1998, 10).

Those three influences—the emergence of conservation biology as a scientific endeavor, specific research efforts in the Northern Rockies, and the efforts of advocacy organizations—are the principal influences that Locke credits with the birth of the Y2Y concept. Yet they are not alone: The zeitgeist of the conservation movement in the early 1990s held several other trends toward landscape-scale conservation, and Locke's thinking was likely influenced by most if not all of them.

For instance, during the late 1980s, a handful of people had begun to emphasize the need for coordinated land management in the Crown of the Continent Ecosystem (CCE). The phrase "Crown of the Continent" had been first coined in 1901 by early U.S. conservationist George Bird Grinnell for the area that would become Glacier National Park, but had since been extended to reach from the Blackfoot River in Montana (near

Missoula) north to Kananaskis Country in Alberta (just south of Banff National Park) (Waldt 2004). A strong proponent of the CCE was Glacier's superintendent Gil Lusk, who had come to the park in the mid-1980s after five years as superintendent of Texas's Big Bend National Park, where he had been recognized by the NPS for his innovative and relatively successful transborder cooperative efforts with Mexican officials and scientists from the neighboring states of Chihuahua and Coahuila (Sifford and Chester, forthcoming; Weisman and Dusard 1986, 63–65). At Glacier, Lusk (pers. comm.) found that U.S. biologists were focusing their efforts on a region commonly called the northern continental divide ecosystem (NCDE), a geographic designation encompassing Glacier National Park and several protected areas to the south (principally the Bob Marshall Wilderness Area) that remains in common use in the United States today (see, e.g., Mace 2004). Hoping to expand their sights beyond the United States, Lusk promoted the broader transborder CCE in multiple venues. However, he recalled that many of the public land agencies—as well as many conservation groups and public citizens—were opposed to the CCE in as much as "all they could see was a federal employee wanting to take over Montana."

Nonetheless, Lusk's promotion of the CCE did reach some receptive ears, at least at a talk he gave at a 1990 CPAWS meeting in Waterton Lakes National Park. Present at that meeting was Locke, who would later recall Lusk as "one of the early leaders in talking publicly about the need to think big and across borders at a scale larger than any individual park complex." Locke added that Lusk had been thinking about protecting landscapes even larger than the CCE—specifically, all the way from Yellowstone north to Jasper National Park. Lusk, in turn, recalled that Locke had been particularly enthusiastic over the CCE concept. Notably, although Lusk did not recall anybody uttering the phrase "Yellowstone to Yukon" at that time, he did recall that some conservationists were thinking as far out as the Yukon (Lusk, pers. comm.).

Also at that 1990 CPAWS meeting was conservationist Beth Russell-Towe, who made a presentation on an international wildlife tourism venture entitled Trail of the Great Bear (TGB, originally Trail of the Big Bear). Russell-Towe (pers. comm.) would trace the roots of the TGB to 1985, describing it as "the first tourism initiative in North America to recognize an integrated ecosystem as a tourism attraction." Grizzly bear biologist Charles Jonkel, founder and director of the Great Bear Foundation, told me that he, Russell-Towe, biologist Charlie Russell, archaeologist Brian Reeves, and tourism marketing consultant Marie Grant had earlier that year begun

> a discussion with Waterton, Alberta, business people who wanted to develop tourism based on a waterslide in the town site. It turned out they merely needed a bit more income, not a waterslide specifically. From that, a group was formed to develop ecologically and culturally sensitive tourism, to not necessarily *attract* tourists but rather to slow them down, to teach them how to enjoy the sidelights and the side roads, and to stay longer, learning how to become part of the land, how to care more—while the local businesses would earn the extra money they needed.

A 1990 assessment of the TGB proposal envisioned it as a "model in ecotourism" that would span "an international scenic corridor" from Yellowstone to Banff (Pannell, Kerr & Foster 1990, 1, 6). The assessment listed major tourism themes for the area as "wildlands and wildlife, major ecosystems, native, historic, cultural and settlement themes," and emphasized that the TGB was "unique" in tourism potential because the region offered both relatively accessible wildlands as well as a high level of tourism infrastructure (Pannell, Kerr & Foster 1990, 2). Today, the TGB is a consortium of more than 350 public-sector and small to medium-sized tourism operations promoting "the well-being of the region's habitats and communities" (TGB 1998). TGB features Y2Y in its literature (TGB 2004–05a, 2004–05b), and Russell-Towe, the president of the company that owns TGB, sits on the Y2Y board.

While TGB focused on an expansive region crossing the U.S.-Canadian border, geographically it stopped near the north end of Jasper National Park. North of there, a group of biologists and conservationists were envisioning conservation strategies to protect this relatively unknown portion of the Northern Rockies. In the early 1990s, biologist Rick Zammuto of the Save-the-Cedar League had conceptualized several "unfragmented biodiversity ecosystem coalitions" (UBECs), each of which would cover a specified region within British Columbia (Sawchuk 1992; STCL 1992). Two of these UBECs were (1) the Headwaters UBEC (HUBEC), which covered the headwaters of the Columbia, Fraser, Peace, and Thompson Rivers in B.C., just southwest of where the straight-line north-south border between B.C. and Alberta begins its jagged route toward the southeast; and (2) the Northern UBEC

(NUBEC), mostly consisting of the Muskwa-Kechika region. "By the end of 1992," Zammuto told me, "there were thirty-six groups in B.C. and some from Alberta, that joined HUBEC." He further pointed out that combined HUBEC and NUBEC constituted "about 80 percent of the Y2Y within B.C." However, although conservationists continued to work on large-landscape conservation initiatives in this area, the UBEC designation never became widely used.

Down south at the U.S. end of what would become the Y2Y region, numerous prominent conservationists were also advocating for large-landscape approaches to conservation. Jonkel, who since the early 1970s had been studying transborder grizzly bear populations with the University of Montana's Border Grizzly Project, recalled that he and Howie Wolke (one of the founders of Earth First!) had coined the phrase "Wild Rockies" at some point in the late 1980s. "The multi-national, dual provincial, multi-state area of the Southern Canadian and Northern American Rockies needs a name," Jonkel (1988–89) wrote, and "Wild Rockies" bestowed upon the region the "geographical and ecological identity" he believed it needed.

By the time that argument made it into print, Jonkel's call had already met with some success in the formation of two groups that appropriated the phrase: Wild Rockies Earth First! (WREF) and the Alliance for the Wild Rockies (AWR). Those two organizations were the institutional results of what Mary Anne Peine (2000) described as the radicalization and subsequent professionalization of the U.S. western conservation movement, and it is worth a short look at that history.

During the 1970s, environmental organizations in the U.S. "Northern Rockies" generally were "small, insular, and closely paralleled the government agencies." This changed in the 1980s with the arrival of more radical organizations such as Montana Earth First!, which later became WREF in order to adhere "to ecological rather than political boundaries." Jumping full throttle into a number of conservation campaigns in the area, WREF found that its methods conflicted with those of the older, established conservation groups. Whereas the latter endeavored to have between one and two million acres designated as wilderness, the former demanded wilderness designation for much larger areas. Arguing that only large-scale wilderness designations would provide adequate protection, WREF often focused on "killing bad wilderness bills"—an approach that pitted the younger activists against more moderate conservationists who saw the large-area proposals as politically infeasible (Peine 2000, 15–16).

By the late 1980s, this younger conservation community had gradually shifted "away from direct action" and toward a professionalism immersed in "improved understanding of environmental laws" (Peine 2000, 17). A sizable segment of this community banded together as the AWR, "a group whose tactics and ecosystem-oriented wilderness proposals fell somewhere between those of Earth First! and the mainstream groups" (Zakin 1993, 417). Using a legalistic approach to large-landscape protection, AWR's principal focus has been on passage of the federal Northern Rockies Ecosystem Protection Act (NREPA), a bill that has been introduced in the U.S. House of Representatives every Congress since 1992—although in 2005 the bill was renamed as the "Rockies Prosperity Act" (see, e.g., U.S. House 1994; 2005). Promoted as "the first legislation to frame wilderness protection in a bioregional context," NREPA called for the designation of more than 16 million acres of U.S. federal lands as wilderness, for the protection of more than "1800 miles of free-flowing streams," and for the establishment of "a system of biological connecting corridors between the core ecosystems in the region" (Bader 1992, 1998–99; U.S. House 1994). Although at least one conservationist has described NREPA as an important inspiration for Y2Y (Foreman 2004, 161), NREPA's geographic extent was still geographically constrained by the international border—a practical jurisdictional problem that AWR explicitly acknowledged (Bader 1992, 62).

Other conservationists were also "thinking big" at the time, and many of them would come to play important early roles in defining the character of Y2Y. Some of these included Yale University professor Tim Clark of the Northern Rockies Conservation Cooperative, conservationist Louisa Willcox of the Greater Yellowstone Coalition and TWP's board (and now with the Natural Resources Defense Council), and Gary Tabor, a veterinarian by training who would draw up one of the first formal analyses of Y2Y (Tabor 1996). Another key individual was Wayne Sawchuk, to whom Locke gave much credit in originally working out the idea of Yellowstone to Yukon. Working under the Chetwynd Environmental Society in conjunction with conservationist George Smith at CPAWS, Sawchuk had begun working to conserve the Muskwa-Kechika region of northern British Columbia (discussed below). Also deserving of note is the fiery Canadian activist Mike Sawyer. Disillusioned by what he described as the Canadian conservation community's piecemeal approach to land protection, he along with several colleagues formed the aggressive Rocky Mountain Ecosystem Coalition in 1993, just as the Y2Y network was forming. During Y2Y's early years, Sawyer consciously took on the role of provocative gadfly on a number of contentious issues that were celebrated within the Y2Y network (Chester 2003a).

These individuals constitute only a sampling of those who had come to accept that real conservation in the Northern Rockies could only be achieved at the landscape level. Overall, however, the historical evidence supports Tabor's summation that it was Locke who "brilliantly synthesized" Y2Y (1996, 7). To reiterate, in working out the idea of Y2Y, Locke built on a wide variety of conceptual approaches to conservation at a landscape scale—some of which he may not have been either consciously aware or did not recall a decade later. As he repeatedly emphasized, his task was nothing more or less than the putting together of ideas that were already fairly widespread in the conservation and scientific communities of the Northern Rockies. Figure 4.2 summarizes those various threads in a shorthand Y2Y "web of conception." Those factors identified by Locke are in circles; others are in boxes. It is important to emphasize that each thread identified in the figure deserves a rich historical analysis and that this schema should be considered no more than a tentative outline.

Institutional Development of the Y2Y Organization

Most of this chapter examines the character and effects of the "Y2Y vision." Yet at the same time that the vision came to be adopted by the conservation community, an organizational structure—often described as "the Y2Y organization," "Y2Y Inc.," and now the "Y2Y Conservation Initiative"—grew up around the vision, and a short history of that growth is in order.

The 1993 Kananaskis meeting was followed by two years of strategizing and building support for the Y2Y concept, after which a series of six meetings were convened between 1995 and 1997. The minutes from those meetings cover a broad swath of substantive and organizational issues, including the development of a mission statement, hiring of a coordinator, strategizing on fundraising, creating a communications strategy, drafting a natural resource "atlas" (originally referred to as a "Rapid Assessment"), and appointing a "circuit rider" to educate U.S. conservationists about the Y2Y initiative (Johns 1998, 10–13).

Early on, an important point of debate circled around the initiative's title of "Y2Y Biodiversity Strategy." In 1995, a "Y2Y Biodiversity Strategy Communications Plan" (Webb 1995) recommended expunging

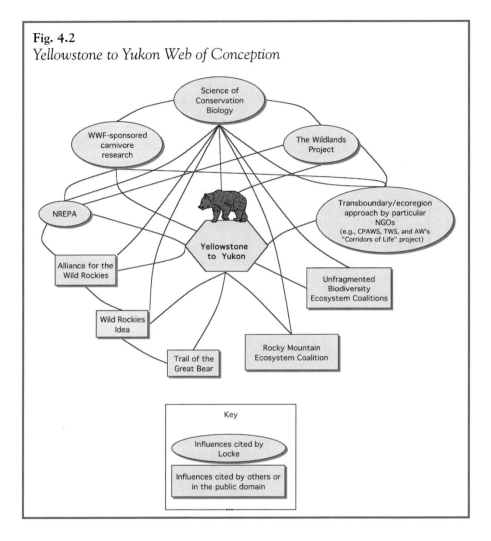

Fig. 4.2
Yellowstone to Yukon Web of Conception

the word "biodiversity" since it "doesn't mean much to most people, but it turns them off." In its place, the report recommended "something that a majority of people will find appealing—not just environmental activists." The reasoning apparently held sway, and over the course of three successive meetings, the name changed from the "Y2Y Biodiversity Strategy" (Y2Y Biodiversity Strategy 1996) to the "Y2Y Network" (Y2Y Network 1997) and finally to the "Y2Y Conservation Initiative" (Y2Y Conservation Initiative 1997b).

In 1996, Y2Y participants were grappling with two key issues: (1) how to structure Y2Y's decision-making authority, and (2) exactly who would be invited to directly participate in Y2Y's development. In regard to the first issue, the internal structure and decision-making procedures of Y2Y were largely laid down at a seminal April 1996 meeting, again in Kananaskis country. While there was "general agreement" to avoid an "overly bureaucratic or controlling structure," many argued for a "modest steering committee, supported by a secretariat, that might have separate Canadian and U.S. components, that would be responsible for communications and for empowering (rather than displacing) grassroots activists." Participants reached consensus that "the structure needs to be 'permeable,' i.e., it must be deferential to local and regional groups and allow others to come or leave as it suits their goals.... The primary focus of the umbrella structure is the large-scale Y2Y effort and coordination within that scale, but not local or regional issues." More specifically, it was agreed to refer to all individuals and organizations participating in Y2Y as "the Council" and to form a smaller "Coordinating Committee" composed of a largely self-selected group of active participants from within the Council (Y2Y Biodiversity Strategy 1996).

Y2Y participants continually discussed the issue of who could and should be involved in Y2Y—specifically, whether the Y2Y structure would "encompass only 'environmentalists' or would reach out to include government, industry, and other stakeholders" (Y2Y Biodiversity Strategy 1996). While some argued that Y2Y needed "an inclusive structure that invites anyone or group to join the Y2Y vision" (Y2Y Biodiversity Strategy 1996), from its conception Y2Y was undeniably shaping up as a networking forum for conservationists and scientists. This became abundantly clear a few months after that formative 1996 meeting when, in October of 1997, Y2Y sponsored "Connections: The First Conference of the Yellowstone to Yukon Conservation Initiative." Held in Waterton Lakes National Park, the conference hosted approximately three hundred people, the majority of whom represented NGOs and joined Y2Y as "Council members." Attendees heard scientists and conservationists present, discuss, and debate the Y2Y concept and review an "interim summary draft" of a "Y2Y Atlas" of the natural resources of the region (Y2Y Conservation Initiative 1997a). A number of media representatives also attended the conference, which generated a flurry of local and national news stories on the Y2Y concept (see, for example, Mitchell 1997; Schneider 1997). Locke recalled that the conference organizers "had to turn away dozens and dozens of people who were dying to come."

With the blessing of the participants at the "Connections" conference, Y2Y had settled into three primary bodies: the Council, the Coordinating Committee (CC), and the staff. From 1997 to the present, the Council has generally met once a year. "When it originated," said conservationist Wendy Francis, the Council was a "voluntary collective of representatives of environmental organizations that included all those who were willing to come together to talk about how the Y2Y initiative as a whole should move forward." While the Council primarily comprised NGOs, several businesses were also enlisted.

The CC was a smaller, self-selected group of Council members who took on the responsibilities of putting a small staff in place and of implementing four key tasks: (1) developing the vision, mission, goals, and principles of Y2Y; (2) strengthening communication networks between conservationists; (3) supporting scientific research in the region; and (4) examining funding needs and sources (Y2Y Biodiversity Strategy 1996). Meeting about three times a year after the 1997 conference, the CC became a close-knit group despite contentious debate that often flared up either in the meetings or over the Y2Y electronic mailing list (although the conflicts were usually between individual members who were already antagonistic toward one another). That the flare-ups could be strong is not surprising: CC members were typically in the position of executive or program director of an advocacy organization—a personality type often lacking in reticence and reserve. At the same time, different CC members brought and shared very different skills and knowledge to the Y2Y table. As but a sampling, the early days of the CC included the following individual members:

- Louisa Willcox of the Sierra Club Ecosystems Grizzly Bear Project brought a practical background of U.S. grizzly bear politics;
- Craig Stewart of the Miistakis Institute brought an applied knowledge of computer and Internet technology;
- Mike Sawyer of the Rocky Mountain Ecosystem Coalition brought an extensive background in energy development;
- Ray Rasker of the Sonoran Institute brought a focus on regional economics and sustainable communities;
- Troy Merrill of the LTB Institute of Landscape Ecology brought a scientist's "check" on how conservationists on the CC understood and relied on both grizzly bear biology and the use of science in general;

- Anne Levesque of the East Kootenay Environmental Society (now WildSight) brought the perspective of community advocacy in B.C.;
- Stephen Legault of the Alberta Wilderness Association and WildCanada.net brought a long background in conservation advocacy in Alberta;
- David Johns of Portland State University and the Wildlands Project brought both a political science and a legal perspective, as well as knowledge of organizational management;
- Wendy Francis of CPAWS brought a mix of legal and advocacy skills, as well as a willingness to take on much of the organizational labor needed to sustain the Y2Y organization; and
- Bob Ekey of the Wilderness Society brought a background in public relations.

A number of factors pushed the CC toward greater institutional stability. One was simply the frequency of the CC meetings. But more important was a working consensus—although it might not have seemed that way to any particular CC member at the time—over issues of "process." This was particularly so in regard to the highly active committees, which included the Operations Committee, the Conservation Science and Planning Committee, the Outreach and Communications Committee, and the Development and Fundraising Committee. As the CC became more institutionalized through the work of these committees, the Council became less an authority-delegating body and, as Francis put it, more of "our word for an annual gathering held in conjunction with the CC meeting each October, where we invite local activists to bring the CC up to date on their activities and provide the local community with a chance to meet and interact with the CC and Y2Y staff."

After an extensive strategic planning process that took place over the course of 1999 and was finalized in the spring of 2000, the CC incorporated itself as a formal board (or technically, as two boards—one in each country—although in all practical senses the two have functioned as a unified body). Although Y2Y participants retained the label of CC for sometime after this formal change, by 2003 that label had largely been dropped for the more publicly recognizable term of the "Y2Y board." The change to a formal board also meant a fundamental change in the role of the broader Council, which no longer had any decision-making authority over the Y2Y board and organization (or what some were calling "Y2Y Inc."). Accordingly, the board decided that the label of "Council members" had become inaccurate and officially decided to describe those myriad organizations as "partners."

As measured by participation records, the size of the CC/board consistently hovered around thirty until it deliberately began to downsize itself after 2003. Measuring the growth of the Council/partners is difficult both because of incomplete records and because "membership" consisted of a loose and informal affiliation. As best as I have been able to reconstruct, there were approximately 118 institutional members in 1999, 160 in 2001, 189 in 2004, and 190 in 2005. In addition to institutional members, there have been individual members, which in 2005 numbered an additional three hundred plus.

As the myriad volunteer activities of Y2Y participants expanded, so did its paid staff. Apparently, the first person to receive Y2Y-dedicated funds for her part-time work was Calgary-based CPAWS staffer Anne Lemorande, who, having helped arrange the 1993 Kananaskis meeting, began dedicating a significant portion of her work schedule to Y2Y in 1994. In September 1995, it was determined "that an executive director [and] secretarial support were immediate priorities and that a communications person would be hired as funds allowed" (Y2Y Initiative 1995). For various reasons—the most important being the general desire to keep Y2Y from becoming an overly hierarchical entity—the title of executive director was soon modified to coordinator. In discussing the requirements for the position, the CC ended up listing a somewhat intimidating list of responsibilities, including selling the Y2Y concept, ensuring effective communication among Y2Y participants, acting as an information clearinghouse on local and regional work in the Y2Y region, acting as the "public interface" of Y2Y, acting as a "referral source for public enquiries," providing assistance to the CC, developing a Y2Y Web page (not, it should be remembered, a common skill in 1996), and planning an annual meeting (Y2Y Biodiversity Strategy 1996). Bart Robinson, a founding editor of *Equinox,* Canada's largest environmental magazine, and widely recognized for his calm, diplomatic demeanor, was hired to the position in 1996 (Wilburforce Foundation n.d.). Y2Y's second full-time employee was hired in 1998, and the growth in staff since then has been sizable; as of 2005 it hovered around ten full-time employees and half as many consultants.

Most of this institutional growth has been financially supported by a handful of environmental grantmaking foundations. Such foundations have traditionally supported a lion's portion of funding for conservation activities in the Northern Rockies. Yet in my interviews, Y2Y participants pointed to foundations not only as a principal source of financial support, but also as key networkers between conservationists in the region. No doubt funders are natural "connectors" for the simple reason that they tend to attract a broad swath of conservationists—all of whom in turn endow the funders with a broad sense of what's going on across the region. Two of those funders in particular have played key roles in Y2Y's institutional development. Ted Smith, the executive director of the Henry P. Kendall Foundation, gave an early grant to Y2Y to build up the electronic networking resources that were critical in the growth of the Y2Y network. Characterized as "incredibly strategic" and "very careful with how he puts money and where he puts money," Smith has also been instrumental in "leveraging" other foundations to support Y2Y.

In 1996, the Kendall Foundation published the first substantial report on Y2Y: *Yellowstone-to-Yukon: Canadian Conservation Efforts and a Continental Landscape/Biodiversity Strategy*. Authored by Gary Tabor, a veterinarian with extensive conservation experience in Africa and Latin America, the report both introduced and legitimized Y2Y to the small sector of the foundation community interested in biodiversity conservation. After writing a second major report on Y2Y for the Seattle-based Wilburforce Foundation (Tabor and Soulé 1999), Tabor went on to direct Wilburforce's regional Y2Y funding program that supports numerous science-oriented capacity-building initiatives within the Y2Y region. Tabor has also taken an important leadership role in the Y2Y network through his consistent participation in the Y2Y CC/board. On many occasions, he has played a critical role as a translator of sorts between participants unable to "hear" what others are saying.

Both the Kendall and Wilburforce Foundations continue to support the Y2Y organization's institutional development. As described further on, both have also developed broader funding programs for conservationists throughout the Y2Y region.

From Mission to Vision

"The amazing thing about Y2Y," said Francis, "is that anybody can invoke the Y2Y name in their little neck of the woods, in their valley that they're fighting for." For Y2Y promoters, this has been both a help and a hindrance, resulting in a confusing mélange of dramatically divergent interpretations of Y2Y's mission. These interpretations have ranged from opponents belittling Y2Y as little more than a "land grab" on the one hand, to extremely broad characterizations of Y2Y on the other—such as that made by Parks Canada official Kevin Van Tighem: "One of the beauties of Y2Y is that it offers an integrated vision for the complete diversity of conservation interests that takes in the entire landscape and says, 'We need to have protected areas, multiple use areas, occupied areas—all these things woven together in a way which keeps this landscape functioning.'"

Given Y2Y's disparate potential target audiences—as well as a body of opinionated and individualistic participants in the CC/board—it should come as no surprise that Y2Y's mission has often molded itself to the eye of the beholder. Yet despite various informal incarnations, the official mission statement emanating out of the CC/board has remained remarkably consistent. Probably the most widespread version came in a full-color, fold-out pamphlet published in anticipation of the 1997 "Connections" conference, which described Y2Y's mission as follows:

> [T]o build and maintain a life-sustaining series of core protected reserves and connecting wildlife movement corridors, both of which will be further insulated from the impacts of industrial development by transition zones. Existing national, state and provincial parks and wilderness areas will anchor the system, while the creation of new protected areas and the conservation and restoration of critical segments of ecosystems will provide the cores, corridors and transition zones needed to complete it. (Robinson 1997)

While this passage shows that Y2Y's roots were in the science of conservation biology, the pamphlet also set out a number of "guiding principles," one of which pointed to the need to promote and encourage "sustainable human communities." By 2001, the concept of community sustainability had become more directly incorporated into Y2Y's public mission statement:

> Combining science and stewardship, we seek to ensure that the world-renowned wilderness, wildlife, native plants, and natural processes of the Yellowstone to Yukon region continue to function as an interconnected web of life, capable of supporting all of the natural and human communities that reside within it, for now and for future generations. (Y2Y Conservation Initiative 2001)

That mission statement remained unchanged through to the present—with the one significant exception that its title was changed from "Our Mission" to "The Y2Y Vision" (Y2Y Conservation Initiative 2004).

Much of the rest of this chapter focuses on the effects of the Y2Y vision. For now it is sufficient to note that—with only a limited number of qualified exceptions—Y2Y participants pointed to the Y2Y vision to justify all the volunteer work they had put into the initiative. The power of the Y2Y vision lay in its positive message, something with which conservationists in the Northern Rockies were not very familiar. As Michael Scott of the Greater Yellowstone Coalition put it, Y2Y offered conservationists the unfamiliar alternative of "an affirmative vision"—which lay in direct contrast to their workaday world of "stopping bad things." And not only was it a positive message, it was a clear and uncomplicated message. Kevin Van Tighem asserted that,

> What conservation needs is a vision in the sense of landscape concepts that people can wrap their minds around—conservation concepts that mean something to them, and that offer some kind of positive relationship between human and nature. And Y2Y embodies all of those—the idea of the grizzly bear or a wolf being able to move safely through the Rocky Mountains without getting creamed or shot—the idea of a future where people can experience the best of the wild, not regret its passing.

Similarly, Y2Y's first coordinator, Bart Robinson, argued that the Y2Y vision is an "idea that's been looking for a home for the last thirty years, that's finally manifested itself in a way and a place that captures the public's imagination." That, he concluded, "might well be the greatest achievement of this whole thing."

Communicating the Y2Y Vision

From Locke's initial conception, Y2Y participants have continually debated the best way to bring Y2Y to the broader conservation community and the general public. In 1995, the tangible results of that debate came in the development of the aforementioned "Yellowstone to Yukon Biodiversity Strategy Communications Plan," which identified environmentalists in the Rocky Mountain provinces and states as the primary constituents of the Y2Y external communications strategy (Webb 1995).

Because this audience would be doing the grassroots work—"writing letters, meeting with local government reps, arguing with opponents"—the conservation community of the Northern Rockies warranted "special attention and a specific style of communication" that included slide shows, tours of critical areas, and seminars (Webb 1995).

The communications plan further asserted that communicating the underlying scientific concepts to the environmental community would be necessary "before we start to win the hearts and minds of middle America/Canada" (Webb 1995). At the same time, however, "securing wilderness habitat" would require broad public support—and the first step toward that would be a "compelling picture" of the need for wilderness protection. Accordingly, the plan recommended several key public messages related primarily to the concepts of extinction (e.g., "Status quo = extinction" and "Extinction is forever") and to the need for large protected areas ("Bigger is better in wilderness protection") (Webb 1995).

By 1998, a "Y2Y Workplan" (Y2Y Conservation Initiative 1998, 3) had identified five key constituencies that Y2Y needed to engage in order to further its mission:

- the conservation community, including hunting and fishing organizations and clubs, single-species advocates, and friendly ranchers;
- land-use managers and decision makers, including those in the First Nations and Native American communities;
- educators and interpreters in the national, provincial, and state parks;
- the academic and scientific communities; and
- the general public.

To gain the support of these constituencies, the Y2Y coordinator's office hired Peter Aengst, who had been closely involved in the successful effort to block a large mine on the perimeter of Yellowstone National Park, as outreach coordinator in 1998. An additional outreach coordinator, Katie Deuel, was hired the following year to concentrate on the U.S. side of the border. Their main tasks were to work with individual Y2Y participants on outreach strategies, to conduct capacity-building workshops, to develop a strategic communications plan, and to prepare Y2Y presentations and publications (Y2Y Conservation Initiative 1998).

LANDSCAPE VISION AND THE Y2Y CONSERVATION INITIATIVE 163

Since 1997, CC/board members, staff, and consultants have produced numerous publications and "innovative products and tools designed to serve the conservation needs of its network cooperators and the needs of the region" (Robinson 1997). These have included:

- *Connections*, a newsletter;
- *Y2Y Conservation News*, a daily compilation of media reports;
- a number of pamphlets, including a sixteen-page pamphlet entitled *Y2Y Q&A 2001*, which contained seventeen questions and answers about the region and the Conservation Initiative (Y2Y Conservation Initiative c.2001b), a handout entitled *The Yellowstone to Yukon Conservation Initiative: Fiction vs. Fact* (Y2Y Conservation Initiative n.d.-b), and two full-color glossy brochures (Robinson 1997; Y2Y Conservation Initiative c.2001a);
- an Interpreter's Kit for use by naturalists and protected area interpreters, which included the Y2Y Q&A booklet, a Y2Y fact sheet, two laminated maps (one of the Y2Y region, the other of the movements of Pluie the wolf), six slides and text, a sheet entitled "What You Can Do to Help Y2Y," a "Bears of Banff Inbreeding Game," a CD-ROM with images of the Y2Y region and conservation biology principles, and a *New York Times* article on "islandization" and extinctions by author David Quammen (Y2Y Conservation Initiative n.d.-c);
- annual reports with overviews of Y2Y activities and finances; and
- numerous scientific publications.

Of all the materials generated by the CC, staff, and consultants, probably the most widely cited report on Y2Y was written by two economists and published by the Wilderness Society. The report argued that

> among the varied reasons for economic growth one of the significant determinants is people's decision to live in a particular place because of its quality of life, including a healthy environment. For the Y2Y region, the natural environment is therefore a stimulus to growth and environmental protection is good for business. It keeps existing businesses from leaving and attracts newcomers, including retirees and entrepreneurs. The environment is an asset that helps

diversify the economy and insulate it from the boom-and-bust cycles of the past. Because the environment plays a significant role in development, land uses that damage the environment, whether a mine, a clear-cut timber harvest or urban sprawl, actually weaken the economy in the long run. (Rasker and Alexander 1997)

The report, *The New Challenge*, soon became widely summarized and cited in Y2Y presentations and literature, and Rasker published the results in several other forums (see, for example, Rasker 1997, 1999; Rasker and Alexander 2000, n.d.). That analysis would also become a focal point for opponents to Y2Y (Chester 2003b).

Probably the most visible public manifestation of Y2Y came from the efforts of Parks Canada employee Karsten Heuer. Between June 1998 and September 1999, Heuer conducted a "Y2Y Hike" from south to north through the Y2Y region, stopping at cities and towns along the way to make presentations on Y2Y. The Y2Y Hike generated significant coverage in the national press and local newspapers and was featured on National Public Radio and the Canadian Broadcasting Corporation, as well as in several national environmental magazines (Burke 1997; Ellis 1997; *Explore Magazine* 1998; Finkel 1999; Heuer 2000; Hutton 1998; Lowey 1997; McMillon 1998a, 1998b; Newquist 2000; Struzik 1998, 1999; Warner 1998). While these articles took different angles on the Y2Y Hike, Struzik (1999) echoed their essential message: "Heuer is convinced the initiative's goal is not beyond reach. Apart from a 14-day stretch in the more developed regions of Montana, he saw signs of grizzlies every day and no shortage of wolf tracks between Banff and Jasper. 'To me, it means some of these larger animals are still able to move about relatively freely,' says Heuer. 'We really need to maintain those corridors that exist now and perhaps restore others.'" Heuer's book on his walk, *Walking the Big Wild: From Yellowstone to the Yukon on the Grizzly Bears' Trail*, has been published in Canadian and U.S. editions (Heuer 2002; 2004).

Expanding the Scale of Conservation for Conservationists

Y2Y has changed the way many people think about wildlife and wilderness in the Northern Rockies, and there are strong indications that that shift has had tangible effects on conservation in the region. No doubt, conclusive evidence to support such claims can be hard to come by—but that is not surprising. Finding causal links between ideas and action constitutes a central problem for academicians of all stripes. But the question is, of course, more than academic, and most Y2Y participants have given some thought—and sometimes considerable thought—as to whether and how an *idea* can actually improve "on-the-ground" conservation in the Northern Rockies. Still, when I asked Y2Y participants about the effectiveness of Y2Y, the typical reaction was a facial cringe. "How am I going to answer this simplistic question," they seemed to be thinking, "when the answer is so complex, nuanced, and subtle?" More specifically, they felt at pains to describe the "effectiveness of Y2Y" for at least three reasons: (1) Y2Y's main effect has been a conceptual shift about land and habitat; (2) the tangible results of Y2Y have manifested themselves through that conceptual shift in a nonlinear manner that is extremely difficult to trace from A to B; and (3) the conceptual shift has occurred at different rates among three broad audiences: the conservation community, government land managers, and the general public. Here I focus on the former, the conservation community, where Y2Y has undoubtedly had its greatest impact (the latter two will be examined in the following section).

Conservationist Stephen Legault, who became involved in Y2Y while affiliated with the Alberta Wilderness Association, answered the question of effectiveness through the lens of the "tipping point" theory, which revolves around the phenomena of "social epidemics" in which ideas "spread just like viruses do"—albeit not necessarily so nefarious as the flu (Gladwell 2000, 7, 19). "A couple of people come up with an idea in the early nineties," said Legault, "and a few years later Canada's most popular broadcaster [David Suzuki] does a television show on it. Well, that's a sign of a social epidemic: influential people contacting more influential people contacting the general populace."

The tipping point theory rests on three "rules" of social epidemics: (1) stickiness, a "simple way to package information that, under the right circumstances, can make it irresistible"; (2) the power of context, or the common-sense (if not tautological) dictum that "little things" can make a big difference when the larger ambient context becomes poised for such a change; and (3) the law of the few, or the idea that social epidemics depend on "the efforts of a handful of exceptional people" (Gladwell 2000). Not surprisingly, the tipping point theory has been criticized by some as "common sense dressed up as science" (Wolf 2000), and at a minimum the three "rules" are more accurately labeled as "relevant factors in some social epidemics" (Sunstein 2000). Regardless of whether "the tipping point" works as a holistic theory of causation, however, each of these three "relevant factors" will indeed apply to Y2Y. But rather than a quixotic quest for some particular tipping point for the Y2Y vision, the following examination is guided by a more nuanced question: To what degree have conservationists adopted the Y2Y vision in their efforts to create a positive "tipping point" for conservation in the Y2Y region?

Before Y2Y became, as Bart Robinson put it, "a force to be reckoned with in a very short period of time," many had been wary about the idea. Robinson recalled a time when "Y2Y was simply an idea that seemed pretty far out there on the edges of things. Further, to the degree we had an identity, it was largely one of stereotypical labels applied by some ultra-conservative media that had discovered us. The *Calgary Herald*, for example, kept referring to conservationists as 'hairy extremists.' As a result, some allies, such as World Wildlife Fund–Canada, were reluctant to be publicly associated with Y2Y since we were seen as a potential political liability."

But within the conservation world, this perception changed rapidly, and nowhere was the conceptual shift more evident than in the way Y2Y expanded conservationists' perceptions of the broader region—a "mindset change in the environmental community," as Kevin Van Tighem described it. For when I asked Y2Y participants to describe how things have changed, they almost universally agreed that conservationists would be thinking on a smaller geographic scale without Y2Y. This potential role for Y2Y was articulated early on by Gary Tabor (1996, 28), who wrote that the "strength of the Y2Y concept" lay in providing a larger context to local and regional conservation activities throughout the Northern Rockies. "Local conservation efforts benefit," he wrote, "when viewed within a broader ecological context." Many Y2Y participants I spoke with echoed Tabor's prognostications. For instance, Rob Ament, the former Executive Director of American Wildlands, described the "power of Y2Y" as providing "a much larger matrix" for smaller conservation organizations "that have never thought of the context of their geographical focus area." Similarly, Ament's successor, Tom Skeele (formerly with the Predator Conservation Alliance), noted that conservationists were getting the message that "Yellowstone is not enough, that Banff and Jasper is not enough." By looking at Y2Y, he said, somebody in Yellowstone realizes all of a sudden that the Crown of the Continent is very important. "Do you like wolves down here?" Skeele then mused in reference to Yellowstone. "Well, we're either going to spend a hell of a lot of money continuing to ship them in or they can walk down here on their own because they made it through the Crown of the Continent."

Conservationist Brian Peck vividly described how Y2Y created genuine excitement among conservationists in terms of fitting the smaller scale into the larger picture: "You kind of get them jazzed up because they go, 'Hey, you know that little 500 acres we're working on? Look where it is, look at this big plan, look where we are; right where Y2Y necks in and gets real skinny—there's our 500 acres.' It's not just their 500 acres anymore; it's part of a 2,000-mile-long core area, buffer, and linkage—and they've got a piece of it. Raises up the ante." Overall, said Peck, Y2Y gets conservationists to expand their working definition of "big landscapes." Without Y2Y, he said, "We would still be back thinking a big landscape was Greater Yellowstone. Which it is—but is it a big, functional landscape the way nature runs the world? No, it's not." In effect, said Peck, Y2Y has

> changed the whole climate of the debate. So now we're not talking just about, well, the boundary of Glacier National Park should be a little bit bigger or, well, we should have another thousand acres here. Now we're talking across the boundary, which we never did before, and we're talking about really big landscapes—2,000 miles of landscapes that need to be looked at. We're talking at the level nature functions at as opposed to tiny little time frames that people work at and tiny little ecosystems that we protect as long as we can't figure out how to get another filthy dollar bill out of them. That's been our history; we look at things that are too small and we think too short term. And Y2Y has completely changed that whole thinking.

From a Canadian perspective, Legault added that Y2Y had created a context for both "our small, local, or regional fights" and "our big international fights, like soft soft-wood lumber agreements and the international endangered species movement." Endowed with this broader context, he said, Canadian conservationists had become better positioned to take tougher stances on conservation issues.

Despite widespread support for Y2Y, it is important to point out that hardly all conservationists in the region have come to work under its aegis. For several years, Y2Y found itself in a contentious relationship with the Missoula-based Alliance for the Wild Rockies (AWR). As recounted above, AWR was formed in 1988 as a coalition of grassroots activist organizations mostly located in the U.S. Northern Rockies (with a few in Canada and outside the region) (Alliance for the Wild Rockies 1995, 2001). While individuals familiar with the origins and consequences of that tension were generally unwilling to have the story go "on the record," the tension resulted partly from personality conflict and partly from division over the appropriate policy stance on the Northern Rockies Ecosystem Protection Act. Nevertheless, over the past few years, the relationship between AWR and Y2Y has significantly improved. Overall, Robinson asserted, any recalcitrance on the part of conservationists to adopt Y2Y "has changed to a considerable degree."

Geographic Predecessors of Y2Y

Several Y2Y participants relied on analogy to describe the rise of Y2Y by comparing it to the conceptual expansion of Yellowstone from a "national park" to a "greater ecosystem," and pointing out the ramifications that expansion had on land management. The origins of that shift are generally traced to the well-known grizzly bear biologists Frank and John Craighead, who began tracking grizzlies in the park and its surrounding national forests in 1959. Often credited with "bringing conservation into America's living room" via their engaging *National Geographic* articles and TV specials (Wuerthner 2001b), the twin brothers and their research team had by 1974 proposed a Greater Yellowstone Ecosystem (GYE) of 5 million acres—just larger than double the size of YNP at 2.2 million acres (Craighead 1979, 4, 239; Robbins 1993, 174). Although the Craigheads were not the first to make such a proposal (see Wuerthner 2001b), Peck pointed out that up to that point nobody had been able to make it stick. But with the Craigheads, he said, "Boom—it becomes the thing to talk about, and all of a sudden everybody talks about the Greater Yellowstone Ecosystem—and it's redefined the debate." And because contemporary delineation of the Greater Yellowstone Ecosystem encompasses approximately 18 million acres—more than six times the size of the park—that redefinition has had deep reverberations (Bengeyfield 2003).

John Varley, director of the Yellowstone Center for Resources (the park's scientific research arm), traced broad public and political acceptance of "the Greater Yellowstone Ecosystem" to congressional hearings on the subject.

He argued that many conservationists in the region simply did not understand the victory they had achieved in merely spurring on the hearings:

> The conservation community is still arguing over whether that was a good idea or not, but there's no question in my mind that it was a good idea because it put it on the map. While some in the conservation community would argue against it because it had no result at the time—there wasn't a product at the end, and so it was widely perceived as being a failure—I argue that it wasn't a failure at all, because what it did was to double or triple or quadruple the members of the public who were aware of it and said, "Why, yeah, that sort of makes sense."

Ken Barrett of the Theodore Roosevelt Conservation Partnership also recalled the significance of the shift. As a staffer at the Greater Yellowstone Coalition (GYC), one of his greatest challenges had been to convince the agencies to use the term "Greater Yellowstone *Ecosystem*"—indeed, he found it hard enough to get them to use "greater Yellowstone *area*." But when the agencies did finally adopt it, the window of opportunity for GYC's advocacy expanded accordingly. The same, argued Barrett, will be true for Y2Y. "When you have an idea as big as Y2Y and as challenging as Y2Y, just getting name recognition and getting the concept into the public arena is a very large undertaking," he said. "The idea that there is an idea—and that that idea is getting greater recognition by both friend and foe—may in fact be the most important thing that Y2Y people have done so far."

It is also interesting to note the institutional similarities between GYC and the Y2Y organization based in Canmore, Alberta. While GYC now operates as an autonomous entity based in Bozeman, Montana, it began as—and to a large extent still retains many of the elements of—a coalition of numerous smaller organizations in the region. Y2Y has followed the same basic trajectory.

In addition to "Greater Yellowstone," Peck also pointed to the Crown of the Continent Ecosystem (CCE) that Glacier superintendent Gil Lusk had promoted in the late 1980s. As another example of a semi-official administrative designation evolving in tandem with the widespread adoption of a biogeographic name, conservationists and land managers now commonly use "Crown of the Continent" to refer to the 5,088 square kilometer (1,964 square mile) transborder region. The result, Peck argued, has been that land managers "are starting to think across those boundaries" in order to look at the landscape in broader terms. "Not just conservationists, not just tree huggers, but agencies are starting to refer to places as the Crown of the Continent—which they never did before." The same evolution, he asserted, is occurring with Y2Y.

Y2Y beyond the Conservation Community

The Y2Y organization has conducted some polling on public opinion of the Y2Y vision, finding that although voters are "generally unfamiliar" with Y2Y, they largely support it after being read a description (notably, 87 percent of Canadians say they strongly or somewhat support it, significantly higher than 66 percent of U.S. voters) (Y2Y Conservation Initiative 2003a). Nonetheless, many Y2Y participants believed that Y2Y's trajectory in terms of public recognition and opinion was unquestionably upward. As the central staffer at Y2Y for several years, Robinson believed that one of the most important successes of Y2Y was in generating broad interest over the "whole question of what do we need to do to conserve wildlife and wilderness." Manifesting itself in "an incredibly powerful, seductive landscape," he continued, Y2Y rapidly established itself as a distinct geographical entity by

> insinuating itself into the general lexicon to the degree that we now hear agency personnel and academics and others referring to Yellowstone to Yukon as if it was identified as an ecoregion by Mackenzie or Lewis and Clark. It's astonishing to me how quickly it's become embedded in the consciousness of people that are concerned with such things. Our greatest accomplishment has been how successfully we've promoted the importance of landscape connectivity. Its importance is that it represents a very different way of thinking about nature, and we've been able to promote that way of thinking in such a way that it's suddenly showing up in all sorts of discussions that are going on among planners, managers, and decision makers.

Robinson was hardly alone in making this general point about Y2Y's broader impact, and it is worth exploring the diversity of expression other Y2Y participants used in describing the similar theme of Y2Y becoming engrained in people's worldview. Ed Lewis, a former executive director of

GYC and later a strategic planning consultant to Y2Y, perhaps most comprehensively explained that "Y2Y has come up with a compelling name, image, and vision that is capable of garnering the awareness, interest, and engagement of large sectors of the public both within the region and beyond." Similarly, conservationist John Bergenske of WildSight described Y2Y as having created "a persona for the region in the eyes of the broader public." Arguing that "this region just hasn't gotten the attention that it deserves," Barb Cestero of the Sonoran Institute's Bozeman office said that Y2Y's greatest success was to put "this landscape on the radar screen of people outside of the region as an important, ecologically significant place that we need to be concerned about protecting." Y2Y associate director Christine Torgrimson argued that Y2Y has become "inserted into the public agenda," and that it has helped to spread the concepts of cores, corridors, and transition areas beyond the conservation community. Bob Ekey of the Wilderness Society colloquially described Y2Y as getting the concepts of conservation biology and landscape connectivity into "the water supply," while Mike Sawyer noted, "There's no question at least that the level of public and political discussion about Y2Y has dramatically increased."

Although the consistency of these impressions is notable, they are nonetheless impressions of people who support Y2Y. Yet extensive media coverage of Y2Y has included both broad and in-depth coverage through the full spectrum of environmental, mainstream, and conservative media outlets. While I have reviewed both favorable and unfavorable reactions to Y2Y elsewhere (Chester 2003b), suffice it to say that between 1997 and 2000 at least seventeen articles on Y2Y were printed in large-city newspapers outside of the Y2Y region, including the *Washington Post*, the *Los Angeles Times*, and the Toronto-based *Globe and Mail*. Furthermore, Y2Y has been recognized and described in numerous popular magazines, academic journals, conservation-related books, and reports. A thorough treatment came in 2000, when the National Geographic Society released the book *Yellowstone to Yukon* as part of its popular National Geographic Destinations series (Chadwick 2000; see also Chester 2001). In 1997, Y2Y was featured on the popular Canadian television show, *The Nature of Things*, hosted by the well-known scientist David Suzuki. Many of the Canadian Y2Y participants would describe that single episode as the single most important event in bringing Y2Y to a broad public audience in their country. Internationally, the IUCN World Commission on Protected Areas has recognized Y2Y as "one of the leading conservation efforts in the world"

(Bennett 2004; Newquist 2000). Y2Y has even merited attention on the popular TV show *The West Wing* and been featured in clothing and biomedical catalogs (Levesque 2000, 163; New England BioLabs, Inc. 2004).

Y2Y's positive coverage in the media likely helped to bring about official recognition of the initiative in both countries. Several Y2Y participants pointed to hard evidence that Y2Y has made inroads into the halls of governments—be they within park headquarters or under capitol domes—and that such acceptance is more prevalent in Canada than in the United States. At a local level, Parks Canada employee Kevin Van Tighem noted that the Y2Y vision had become "a very important tool for me to position some of our issues, and positioning issues is important in a sense that that's the way you generate management support for doing stuff—that's how you generate resources and mobilize human energies." In 1999, Parks Canada displayed a map and description of the Y2Y region, the Y2Y initiative, and Heuer's Y2Y Hike in its annual visitor's guide to the Rocky Mountain National Parks of Canada (Parks Canada 1999). The following year, a "Panel on the Ecological Integrity of Canada's National Parks," which had been commissioned in 1998 by the minister of Canadian heritage to provide recommendations for maintaining the parks' ecological integrity, stated that conservation at the scale of Y2Y "is the new paradigm of protected areas—from islands to networks" (PEICNP 2000). Y2Y has also been positively referred to in the provincial legislatures of Alberta and British Columbia (Hansard 1999a, 1999b), and in June of 2000, the minister of Canadian heritage referred to Y2Y while testifying before a Senate committee for a new national parks act (which passed in October of the same year). "You'll never read it in the text" of that act, asserted Legault, but Y2Y "is behind that text."

In contrast to Canada's recognition of Y2Y, participants agreed that Y2Y has had less influence on the U.S. side. However, reflecting the comments of several other Y2Y participants from the United States, biologist Lance Craighead believed that although U.S. government officials have not widely adopted the term "Yellowstone to Yukon," they are slowly coming to "accept it like part of the thing that they're working toward." This was clearly evident in 1998, when Parks Canada and the U.S. NPS signed a Memorandum of Understanding that identified Y2Y as a "high priority geographic area of potential cooperation" (Sandwith et al. 2001, Appendix 3).

Although recognition from their respective national governments has been a welcome sign of progress to Y2Y participants, garnering governmental involvement in Y2Y has generally not registered as an important objective (although a few Y2Y participants have wondered whether it should). In contrast, Y2Y participants have struggled to involve aboriginal peoples—known as "First Nations" in Canada and "Native Americans" in the United States—in Y2Y. Although Y2Y participants repeatedly raised concern over the lack of aboriginal participation, outreach to aboriginal communities had been largely unorganized and sporadic, with the exception of outreach conducted in a far northern part of the region (as will be covered later in the discussion of the Yukon). In 2002, however, Y2Y held a meeting entitled Exploring Common Ground on the Dena Tuna: A Gathering of First Nations and Native Americans from the Yellowstone to Yukon Region at the townsite in Banff National Park (Y2Y Conservation Initiative 2002). As a region that "has been used as a migration corridor by aboriginal peoples (and animals) for thousands of years," Dena Tuna is "the name given to the mountainous Y2Y region by the Dene people of northern British Columbia and the Yukon" (Y2Y Conservation Initiative 2003c). Wendy Francis, the chair of the Y2Y board at that time, addressed the various aboriginal representatives at the meeting:

> Y2Y wants to learn how to work respectfully with First Nations and Native Americans. We need to look for shared goals and find ways to support your traditional work and lifestyles. We recognize that how each of you related to the land and work with it are different based on your histories and the treaties and constitutional rights you possess (Y2Y Conservation Initiative 2002, 7).

The meeting resulted in two aboriginal representatives joining the Y2Y board, the establishment of a Y2Y Aboriginal Advisory Group, and an updating of Y2Y's "Principles for Working with Aboriginal Peoples" (see Y2Y Conservation Initiative 2005).

In sum, support for the Y2Y idea did not emanate solely out of the conservation community, but also manifested itself in media exposure, in acceptance from segments of the private sector, in implicit approval from certain branches of government, and in improved working relations with First Nations and Native Americans. Although this support did not constitute widespread societal adoption of the Y2Y ideal, there is little doubt that Y2Y had made an important first step toward broad public recognition and acceptance. Why this support arose will soon be examined, but first it is important to look at the antithesis of support for the Y2Y vision.

Vision Backlash

Even as many were welcoming the Y2Y vision, Y2Y's quick rise in the public arena engendered an immediate and vocal opposition to what was labeled "yet another" antihuman crusade by the conservation community to "lock up" public and private lands—this time in the form of an unfathomably vast international wilderness preserve. Much of this opposition to Y2Y was reported in the news media, where numerous editorials and industry reportage have been highly critical of Y2Y. In addition, several local municipalities and Chambers of Commerce publicly repudiated the Y2Y vision. Geographically, perhaps the most vocal opposition to the Y2Y concept has arisen in southeast British Columbia, where Y2Y has had a high profile. And while opposition to Y2Y has been more evident in Canada in general, at least one U.S. county has seen significant opposition to conservationists because of their connection to Y2Y (Chester 2003b).

Y2Y thus symbolized for many a hidden agenda to disrupt development, livelihoods, and a way of life. Why did such strong fears arise? Conservationist and hunter Jim Posewitz, an early "supportive critic" of Y2Y (Posewitz 1998), argued to me that the combination of grandiosity with a lack of details caused the Y2Y vision to backfire:

> Y2Y quickly become a local pariah.... Here's this monster entity going to take the whole continent or at least a great big chunk of it. And so the wise-use people and the Blue Ribbon Coalition [an NGO advocating for motorized access to public lands] felt that they could hang the albatross of an international conspiracy on a local timber sale or local road-closing debate. It created an atmosphere susceptible to picking up local political energy charging that [Y2Y] was part of not only a national conspiracy, but, by God, it's international—next thing, you're going to have the UN in here.

For Y2Y participants, when it came to the more extreme conspiracy theories that were bandied about, silence was often chosen as the best retort. In other cases, Y2Y participants answered with incredulity and humor (much as had many of the ISDA participants in response to similar accusations). For example, in response to an accusation of working under a UN-sponsored land grab, Y2Y participant Lance Craighead (1998) wrote in a letter to the editor of the *Bozeman Chronicle*: "If anyone out there is part of this ecoterrorist paramilitary arm of the United Nations, or is otherwise concerned about environmental issues, please send us a (tax deductible) check to improve our web site. We also need help maintaining an office . . . and the way things are going we may have to find a bigger space to conceal the black helicopters that UNESCO and UNICEF should be sending us soon."

Not all of Y2Y's critics came from the extreme right, however, and Y2Y participants found themselves facing legitimate concerns both over a perceived ambiguity within Y2Y's goals and over the appropriate role for extractive industries in the Y2Y region. To such concerns, Y2Y participants have generally responded that Y2Y is *not* about excluding human occupation of the Northern Rockies, but *is* about ensuring wildlife occupation throughout the region. As such, their public repudiations to their critics in various forums, including community gatherings and local media, were at times more akin to attempts at reconciliation.

Behind Y2Y's public response to its critics, Y2Y participants were spending a significant amount of time in internal deliberations over how to respond to the accusations that were being made about Y2Y. Although it simplifies matters somewhat, it is fair to say that the CC split over how to respond. Some Y2Y participants were sympathetic to the fears, believing that the outcry against Y2Y resulted from poor internal strategy over presenting it to the public—a strategy that had been too hastily built up within the confines of the conservation community and then foisted too quickly onto the public stage without adequate anticipation of the antipathy Y2Y would engender. Other participants believed that backlash was to be expected no matter how Y2Y was presented to the public—and that short-term backlash was even something to be welcomed in terms of garnering attention to the idea.

This split revealed that despite similar fundamental beliefs about the importance of conservation, different Y2Y participants held significantly different viewpoints over what constituted an appropriate and effective public relations strategy for Y2Y. For example, at one point during this broad debate, a specific point of contention was whether conservationists —particularly "outside" conservationists—should even use the phrase "Yellowstone to Yukon" in certain localities where backlash toward conservationists had become divisive. Due to the strong individual personalities who held diverse opinions over the issue (there being few shrinking violets to be found among the advocates participating in Y2Y), the debate over the issue was at times vociferous and contentious. But the fact that there was some conflict over how to portray Y2Y in public should never have been altogether unexpected.

Ultimately, the backlash to Y2Y can be only partly explained by any ambiguity regarding its core messages about wildlife and wildlands. A more likely cause of the backlash was that Y2Y was assembled from powerful imagery recognizable to anyone—not just to those within the self-elected conservation community—who understood the broad sweep of North American history. In other words, opponents early on were likely reacting to the potentially powerful iconic imagery inherent in a movement that incorporated the names "Yellowstone" and "Yukon," a movement that had the potential to change the way society manages land and wildlife across large landscapes—if only it were to become engrained in the broader society. For no placid abstractions along the lines of, say, "wildlife corridors" or "expanded protected areas" or "ecosystem integrity" had elicited a fraction of the negative response to "Yellowstone to Yukon." Consequently, the strong reaction to Y2Y may be attributable to a proactive attempt to cut Y2Y off at its roots before it could make any lasting impacts.

If that were indeed the case—if opposition to Y2Y derived less from lack of clarity over its mission than from the inherent power of the Y2Y vision—it might serve to alleviate the concerns of those who believed that Y2Y had gone too far and too fast in the public arena. Indeed, had the Y2Y message (or rather, the suite of ambiguous Y2Y messages) not been widely disseminated from the start, the counterattack could feasibly have been far more effective in suppressing the mere use of the term "Y2Y."

Y2Y's Iconic Composition

> *I am the land that listens, I am the land that broods;*
> *Steeped in eternal beauty, crystalline waters and woods.*
> *Long have I waited lonely, shunned as a thing accurst,*
> *Monstrous, moody, pathetic, the last of the land and first.*
> —Robert Service

As Y2Y's former conservation science coordinator Marcy Mahr stated, Y2Y was "becoming more of a household word." Yet Mahr emphasized the word "becoming," for at the same time that she and many others made a positive assessment of Y2Y's growing public acceptance, they typically rejoined that Y2Y has a long way to go in that regard—and not a few voiced concern over the relationship between the Y2Y vision and "on-the-ground" conservation success. But accepting the premises that Y2Y (1) has changed the way most conservationists in the Northern Rockies think about where they are and what they are doing, and (2) has at least begun to insinuate itself into the public sphere, the challenge is to examine how this actually happened, and that brings us back to a theme we have already touched on: Y2Y's iconic composition.

In terms of evocative geographical appellations, it would be hard to best the assonant combination of "Yellowstone" with "Yukon." In the United States, the designation of Yellowstone as the world's first national park underpins the expansion of the country's conservation movement and subsequent environmental movement. "Since its establishment," as Judith Meyer (1996, 5) wrote in an extensive cultural history of the park, Yellowstone "has matured into a national and international symbol of our desire to protect and preserve natural landscapes." And in *The Law of the Yukon*, quoted above, poet Robert Service (1953) enshrined the rugged Canadian territory as the proving ground of national grit and identity. Individually, then, both "Yellowstone" and "Yukon" represent something far more than geographic designations; they are as much icons of North American self-perception as they are places on a map.

In incorporating the two into a single phrase, Locke described the choice of words as highly deliberate:

> I chose those words, "Yellowstone to Yukon," because they're deep symbols in people's brains. If I say those words in Stuttgart, Germany, in Toronto, in New York, or in Tokyo, everybody knows what I'm talking about. I'm not selling soap, I'm not selling cars, I'm not selling running shoes. I'm talking about grizzly bears and wolves and wilderness and wild places. Everybody knows that when they hear those words—that's why I picked them. It was very conscious. I could have called it the "Great Rockies thing" or the "Cordillera whatever," or it could have been called a lot of stuff, but these words—Yellowstone to Yukon—work. And they work because of Robert Service, who's part of every North American's cultural story, and he's widely read in other places. The "Call of the Wild," the

Yukon, the remoteness. Yellowstone, the world's first national park, the symbolism. Everybody knows about Old Faithful and cares about nature. That's why those words work. People go "ding" when they hear "Yellowstone" and "Yukon." They're equally powerful symbols in the imagination of the Western world *and* in the case of the Eastern world—in Japan! Very powerful symbols.

Yet even as widespread familiarity with both "Yellowstone" and "Yukon" were critical in generating a strong response to Y2Y, its success within the conservation community was a result of more than the simple combination of two icons. Numerous Y2Y participants repeatedly emphasized that Y2Y has attracted attention because of the expansive and holistic vision it put forth. Robinson, for example, asserted that

> Yellowstone to Yukon resonates with North America's fundamental notions of wilderness and wildlife, and it arrives at a time when people are not only taking the measure of the impacts of 150 years of unfettered development, but are noting that the pace of that development is accelerating. The notion of landscape-scale connectivity is an ameliorating vision and holds out the hope of the continued coexistence of natural and human communities. The idea of connectivity has been floating around for decades, but it took one of the world's most beloved landscapes—and the perception that the native species and natural processes of that landscape are threatened—to bring the idea to ground.

Many other Y2Y participants had similar comments—the collective implication of which was that in order to assess the fundamental effect of Y2Y, it is of primary importance to understand the "grand vision" lying at its core.

"Few, if any forces in human affairs," asserts MIT professor Peter Senge (1990, 206), "are as powerful as shared vision." In his popular book *The Fifth Discipline*, he argues that a shared vision is not an abstract idea—"not even an important idea such as freedom"—nor is it a "solution to a problem" (206, 214). Rather, a shared vision is a "force in people's hearts, a force of impressive power" that answers the question "What do we want to create?" (206). With a vision, he summarizes, people "do not focus on the long term because they have to, but because they want to" (210). Although Senge was writing about business strategies, Y2Y participants keyed into something very similar to Senge's conceptualization of a shared vision when they crafted the aforementioned Y2Y mission statement—which, tellingly, would later be formally renamed "the Y2Y Vision."

Both in conversation and in public presentations, Locke is unwavering in his insistence that the Y2Y vision—or "myth," or "story," as he alternatively put it—has been and will be a critical factor in tangibly changing the way people understand and manage the region's land and wildlife. In order for Y2Y to work, he argues, it must be about "creating a myth, not in the sense of a lie, but in the sense of a new story by which people live." Myths, stories, self-conceptualization, he says,

> that's how stuff really moves in society. This is what I learned in politics. And it doesn't move because Harvey Locke, the president of the Canadian Parks and Wilderness Society, says this is the right thing to do. It moves because that's a symbolic reflection of what people want to have happen and that you're articulating their dreams. So how do you influence their dreams? That's the question that conservation is about and that's the question that's almost never asked. This is about influencing dreams. This is about helping people want what nature needs as their dream. And when you get there, then you move something. And to just fight conservation as a land-use battle, which is how it's always been done, is a loser's game because it's always a game of tradeoffs.

Y2Y is working, Locke said, because it is *not* playing the game of tradeoffs, in which "environmentalists want to save X, developers want Y—we'll saw this thing in half." Rather, on a scale unseen before in North America, Y2Y has been effective "because it's worked at the basic level of story about how people relate to landscape." What Y2Y did, he continued, was to create the intellectual, spiritual, and emotional context—the story—within which people could see that creating large protected areas and connecting corridors was "the right thing to do."

Because Locke is professionally and personally invested in Y2Y's success, there is room for skepticism of his constant focus on the Y2Y vision. When asked to provide tangible examples of Y2Y's on-the-ground conservation achievements, he responds with specific cases and hypotheticals. As but one example, when telling me about the successful pre-Y2Y campaign to block the expansion of the townsite of Banff (located within the national park), he argued that today "you wouldn't even have to do the campaign because of Y2Y—the stuff that was being discussed and we were trying to stop then is inconceivable to discuss today; it wouldn't even be on the table." But even when focusing on specifics, Locke nonetheless always brings the focus back to the vision. Pressed to explain why all roads lead to the vision, Locke cited the novel *Ishmael* by Daniel Quinn (1992), which he described as "the story of a Socratic conversation" between a gorilla and a human. At once quirky and profound, the book has attracted a dedicated following and has been translated into twenty languages (see http://www.ishmael.com and Quinn 2005). Locke's strong affinity for the book results partly from its message that "the world doesn't need to belong to man—but it *does* need man to belong to *it*" (Quinn 1992, 243). But from a more practical perspective, Locke saw the book as a guide to understanding the unconscious foundation of how people understand their role in the world—a foundation that the novel's protagonist calls "the story" (Quinn 1992, 50). As Locke summarized, "Quinn's basic thesis is you go to the story if you want to change the behavior. You don't play on the turf of the story that's being told now and try to fine-tune it, like, 'Let's save the best of the last,' or 'Let's do this because we can afford it.' Uh-uh. You say, 'Let's do this because it's right. Let's do this because it's the world we want to live in. Let's do this because this is our story.'"

Locke also pointed me to the writings of his colleague David Johns, a political scientist and a highly active participant within the Y2Y organization from the first 1993 meeting. Johns has described "stories" as "more than just a means of trying to *describe* the order of the universe, be it physical or social. They are also the primary way we *create* order and meaning. There is a real world governed by regularities, and we can discover those regularities and test the truth of our descriptions. But we must create meaning and purpose as individuals and societies" (Johns 2003a, 172). He goes on to note that even though it would be "ridiculous to assume that we can engineer fundamental beliefs," he argues that it is possible through story to change "lower order beliefs" (2003a, 173–174). Environmental values largely belong to this hierarchically lower category, and he points to some specific examples of how particular stories put forth by segments of the environmental community have changed such beliefs. For instance, he described how the Evangelical Environmental Network derailed efforts to weaken the U.S. Endangered

Species Act. "Arguing that the natural world was God's handiwork," Johns relates, "the network undercut the sacred mantel that developers and despoilers often hide under" (see Barcott 2001).

Locke further described the power of story by pointing to a counter-example to the environmental movement, specifically the "wise use" movement as a prototypical example of people using the power of a vision. The wise use movement, he said, had successfully convinced a large portion of the populace "to hate anything that's international, anything that involves government, and anything that smacks of controls over land use." This was nothing less than a "tremendous success" marked as much by what it had prohibited as by what it had created:

> If you ask me to show you what the wise use movement and its cousin the right wing have accomplished in a statute, I would say, Huh? What? Wrong question. It's the *absence* of many statutes and the *absence* of land-use planning in so many places that they've accomplished. Which is exactly what they wanted to do. In the seventies, people said, "The government should do this and the government should do that." They didn't say, "The government shouldn't exist or the government is bad." The wise use movement and the right changed that in twenty years, and the change has been pervasive. And their success demonstrates the power of a wide-spread idea. And the impact of that idea can be measured anywhere on the landscape in North America—it's the same in Canada and the U.S. That's where the idea succeeds because it's not in the control of any one person, not Newt Gingrich, Preston Manning, or Stockwell Day—nor of George W. Bush or Ronald Reagan. It's much more than that. It's flowing in the social groundwater. That's when you have an impact.

To no small degree, of course, it was that reactionary impact that Locke was trying to counteract with Y2Y. "We need this idea of big landscapes and mammal corridors with big core protected areas," he averred, "to be as strong and as inherently right as the way the right wing was able to make the idea that government is bad and greed is good pervasive in the 1980s. We're doing the same thing the wise use movement did—and we're going to beat them at their game."

Locke carefully added that he is "not so naive as to think that you're going to have everybody of one mind all the time on everything." But he remains convinced that the conservation perspective is becoming preponderant:

> Is there genuine, general agreement virtually everywhere that this is the right scale to think in both those communities? Absolutely there is. Is there support in varying levels in those communities for this idea? Absolutely there is. Is there engagement from people who matter in the region in this stuff? Absolutely there is. I don't know of a better acid test than that.

"The idea being there," he summarized, "everyone has to respond to it." And while Locke generally focused on Y2Y's impact on the broader society, he emphasized that without Y2Y an array of conservation organizations would still be fighting individual campaigns and spending all their time trying to devise a coordinated "battle plan" for conservation. The earth might "burn around you" in the meantime, said Locke, "whereas if everybody is empowered with the bigger scale, big things can happen." The only significant roadblock to implementing the Y2Y vision, he all but stated, is self-imposed fears of asking for too much. "You'd be hard-pressed to find anybody in the region who says this isn't the right scale," he said. "You have people say operationally this is a fantasy, or you can't do that, or the wise use movement doesn't like it. You get told all the reasons why it can't happen, but you never get told it's not the right thing to do. And that's the inherent power of the idea."

From Concept to Conservation

I asked Y2Y participants the following question: What would the region look like if there had never been Y2Y? While their responses ran the gamut from "no tangible difference" to "a different world," most of them landed near the middle leaning toward the latter. Their answers typically incorporated the mindset change brought about by the Y2Y vision. For instance, when I pressed Legault for a tangible example of how the Y2Y vision had had an impact, he responded that it was "extremely helpful" in his work to protect the popular recreational area south of Banff NP known as Kananaskis Country:

> For four years, I was the chair of something called the Kananaskis Coalition, a coalition of thirteen recreation, community, and

environmental organizations. One of our mantras was "This is going to be good for Y2Y." It helped us understand that by working to protect this 4,200-square-kilometer multiple-use landscape, we were doing something that would ultimately help Yellowstone to Yukon achieve its vision. I know of many activists that I've spoken with up and down the Rockies who feel the same way: They see their work on the ground to stop the Cheviot coal mine, to protect the Bow Valley, to protect the Castle Crown Wilderness, to protect the Crow's Nest Pass as being something that they're doing on behalf of the region to help achieve this vision. And it's given people a hook to get into a much larger vision.

Other participants gave Y2Y credit for having played a significant role in protecting several areas, most notably the Bow Valley (both within and outside of Banff National Park), the Muskwa-Kechika, and the Yukon (the latter two of which are described at length later on). In no case, however, was there unanimity among Y2Y participants regarding the *degree* to which Y2Y had played a role—in fact, their answers were often quite disparate in this regard. Furthermore, with the exception of Locke and a handful of others, Y2Y participants expressed some concern over the opportunity costs of involvement in Y2Y—that is, over whether all the time and resources that they had devoted to Y2Y would have had a greater payoff if devoted to any one of the plethora of place-based conservation campaigns within the Y2Y region.

However, when I moved my questioning away from specific geographic locales toward other potential benefits derived from Y2Y, concern over its opportunity costs significantly dissipated. The resultant list of "conservation successes" I gleaned from Y2Y participants is presented in the following paragraphs. Although not listed in order of priority, it was the first item—Y2Y's role as a nexus for networking—that uniformly loomed largest in their responses.

Networking

Every participant with whom I spoke emphasized Y2Y's critical role in establishing a durable network among conservationists in the Northern Rockies. Even the skeptical Mike Sawyer credited Y2Y for being "very useful—there are groups who are active in the landscape who, prior to Y2Y, no one knew existed, and now they're part of this large family."

While there were many nuances as to what they actually meant by networking, most participants focused on how Y2Y enabled collaboration, or simply brought people together who could only benefit by getting to know one another. The most enthusiastic on this point was Brian Peck:

> We keep talking about Y2Y being a network. Man, is it ever; it's a network where you really get to network. You come to these meetings, and you meet new people all the time. Or someone will grab you by the arm and say, "Are you still working on x, y, z?" And you'll say, "Yeah, I am working on x, y, z." "Wow! Well, this guy who I just talked to over here, he's been working on that on the Rocky Mountain Front—you ought to talk to him." And so there's this constant interaction, and it's almost like this rejuvenation of the excitement, and it's building, building, building all the time.

Several participants focused on how Y2Y had expanded people's understanding of the region's institutional landscape. Robinson argued that if there had been no Y2Y, "you'd still see groups working in isolation on their own local projects—not thinking very much about connectivity, because conservation groups are generally organized to move from brushfire to brushfire. There's very little time for long-range, big-picture planning. Everybody's aware of the need to put what they are doing in some context, but you never really quite get there. Well, that's one of the elements that Y2Y has brought, certainly and solidly, to conservation in the region."

"We know more of the players," summarized Tom Skeele, "and now I know more groups that we can work with." Rob Ament added that the high level of professionalism among Y2Y participants had been key in establishing useful connections between his group (American Wildlands) and others.

Other Y2Y participants focused on how Y2Y engendered broad-based support for specific local conservation campaigns. Arguing that conservationists have learned "to think more strategically and broadly" by working together through Y2Y, Ed Lewis pointed out that their ability "to tap into the network for constituency support on site-specific issues has been utilized again and again." Noting that "we always know that we have some people to rely on if we need input," Ament found Y2Y helpful to American Wildlands in generating activism on particular conservation issues. And Wendy Francis spoke for many in describing how Y2Y "opened up a whole new avenue of campaigning where you could draw on other network groups, send them action alerts, and they would understand why it was important. Suddenly my campaign goals became important to them and vice versa."

Overall, Peter Aengst asserted, by hooking "themselves to the Y2Y vision," conservation organizations in the Northern Rockies "are now more healthy and more savvy than if Y2Y had never existed." Or as Gary Tabor (1996, 28) eloquently described Y2Y's effect (now often quoted in Y2Y literature), "Conservation efforts are like the landscapes that they seek to protect. Just as a weakened, isolated ecosystem is strengthened by connections to other land areas, separate conservation efforts can gain from the support provided by a larger network such as Y2Y."

Since I conducted my formal interviews, the role of networking under Y2Y has changed somewhat. During recent deliberations within the board over Y2Y's changing priorities, for example, several have argued that networking per se should no longer be a "primary goal" of Y2Y. That such a dramatic change would even be a topic of discussion has probably occurred due to one particular outside force. Specifically, the networking function that Y2Y provided in the mid-1990s has been usurped by the Internet—one might even say by Google. For in 1997, Y2Y was a rich resource if you needed to know, say, who were the region's most innovative lynx researchers; but less than a decade later, innovative lynx researchers have become a Google search away. While this is an exaggeration—everything about people who study lynx may not yet be on the Internet—it does mean that conservationists now have a far more expansive range of networking options than they did just ten years ago.

Indeed, the evolution of the networking role of Y2Y runs in rough congruence with the history of Y2Y's online technologies. While I have extensively covered that history elsewhere (Chester 2003a), it is worth summarizing several significant points here. Most important, it was electronic communications that formed the foundation of a durable and widespread Y2Y network. As Kevin Van Tighem noted, when he first heard of Y2Y in the mid-1990s, he thought that it would "fizzle, that it was too large and the interests were too complex." But, he countered himself, "I didn't take into account the power of electronic technology, communications technology." And the functional bedrock of that technology was the Y2Y Council electronic mailing list (or EML; note that most Y2Y participants and staff have commonly referred to the EML as the "listserv," which is the brand-name for one of the first widely used EML software packages). Established in 1995 with 28 subscribers, the EML expanded to 88 subscribers in 1997, 210 in 1999, and 303 as of 2004. It has been used to disseminate a wide range of information, the most prominent types of which have been activist alerts, information requests and responses, and organizational dissemination of information (e.g., agendas, minutes, documents for Y2Y meetings). A number of subsidiary EMLs (e.g., for science and the CC/board) have also been established under Y2Y.

Y2Y participants described the EML as key to Y2Y for four primary reasons:

- it is exceptionally convenient, efficient, and easy to use, particularly compared to phone calls, mail, and faxes;
- it has given Y2Y a constant presence in the daily lives of conservationists, helping to maintain a bond among Y2Y participants;
- it has allowed conservationists to reach a larger audience for activism and funding initiatives; and
- to a greater degree than the Y2Y meetings, it has brought people together who otherwise might never have known about each other.

Yet for all the benefits of the EML, many Y2Y participants pointed out that it has been problematic—for instance, by providing a ready forum for occasional flare-ups of antagonistic or inflammatory language. Overall, however, such incidents over the EML have been rare ("families have squabbles," noted Craig Stewart). A more serious and systematically engrained problem has been that of the overwhelming amount of material coming across the EML. Most participants expressed nothing less than dismay at the level of "information overload" it wrought. It is important to recall that Y2Y was born just before conservationists in the Northern Rockies began to adopt online technologies. Y2Y's EML was a new toy when it was created in 1995, a "gee whiz" technology that was fun to play with and that—critically—benefited from a relative dearth of competing electronic background noise at least during its first few years. But for most if not all conservationists who use the Internet on a regular basis, that noise has today become a constant thunder.

Nonetheless, amid today's unceasing electronic din, it appears that many Y2Y participants remain active followers of the Y2Y EML, perhaps in recollection that it was the principal mechanism responsible for creating a sense of community among conservationists throughout the region (as well as a number of conservationists outside the region). In other words, and this is admittedly conjectural, there seems to be a certain loyalty to the Y2Y EML inasmuch as it was the glue that held together a nascent movement that has since grown into a powerful force.

Inspiration in a "Landscape of Hope"

In addition to expanding the geographic scale of how conservationists think about the Northern Rockies, the Y2Y vision has constituted a wellspring of inspiration for conservationists. Several of them noted that to no small degree the inspiration came from the strong sense of camaraderie that the Y2Y network had engendered among conservationists.

Alternatively, noted Ament tongue-half-in-cheek, the quotidian effect of Y2Y was that it had kept people coming to Y2Y for "seven years of meetings—that's a lot to ask of anybody." But most Y2Y participants circled back to the sheer inspirational effect of the Y2Y vision. Paraphrasing Wallace Stegner, Stephen Legault said that the power of the Y2Y vision was to provide a "landscape of hope" within which conservationists could see where they wanted to go. Y2Y has given people a reason to "wake up in the morning to sit down and get on the phone for 11 or 12 hours a day and be on the computers for 11, 12 hours a day, fighting what can be at times a very boring, mundane, and routine battle. Y2Y has contributed significantly to that landscape of hope by giving a lot of us something to hold on to; it's given us a vision beyond the boundaries of the valley that we live in or the campaign that we're currently working on."

Similarly, Louisa Willcox pointed to the role of Y2Y in keeping her inspired: "It's the right scale to work on, it's the right thing to do, and that's what drives me. Also, I'm personally inspired—which is sometimes hard to feel in some of the day-to-day work, which is much more trench warfare. We're entangled in a long slog, a siege mentality, trying to protect grizzly bear and other wildlife habitat in a place where every acre is fought over."

Bart Robinson argued that the Y2Y vision had been even "more than inspirational" in providing "a grounding for the work itself—once you've seen how those connections can exist, then it really becomes part of how you're thinking and how you relate to the work that you're doing." He recalled that at one of the early Y2Y meetings, a long-term conservation advocate pronounced that she finally understood "why I've been fighting so damn hard in my own backyard all these years—it is part of a bigger picture, and this is what the bigger picture looks like." All of a sudden "there's a context and a meaning to the work that she's been doing locally which wasn't there before."

This inspirational effect of Y2Y, to which CC/board members attributed their dedicated involvement in Y2Y for so many years, has also been widely used by Y2Y participants in their own outreach efforts. "It's captured people's imagination," said Ray Rasker of the Sonoran Institute, "that there is still one place that is so wild that you can think of this scale and dream of as a possibility. It's huge, it's enormous—nobody has come up with anything of this scale ever." This "powerful dream," former Y2Y staffer Peter Aengst further argued, has brought in people who have "never been involved with conservation in all their life—or at least as a member of a conservation group. What I repeatedly found was that the first thing that really turned them on to conservation was the Y2Y vision. Even if they didn't end up doing anything with Y2Y, it often opened up a whole new world for them." In practical terms, Skeele concluded, the attractiveness of the Y2Y vision has played "a significant role in reformatting, redefining how we go about doing conservation because we're going to redefine how we sell wildness and wildlife to people."

Learning across Borders

Before Y2Y, conservationist Ed Lewis noted, transborder networking had been tried "but it hadn't worked very well." The Y2Y vision changed that by "getting Canadians and U.S. folks to think and work across the border." Locke echoed that point, arguing that before Y2Y, "people weren't thinking across the border—they just weren't. Waterton and Glacier a little bit, but not in any kind of a way where they felt that their ideas and interests were legitimate and accepted as legitimate in a transboundary sense."

For many participants, Y2Y opened the intellectual border between the two countries, providing a critical international learning forum. There were two primary types of transborder learning: cultural and strategic. In the first case, Y2Y provided a forum for U.S. conservationists to learn about Canada and vice versa—although the latter to a lesser degree, since Canadians have a comparatively stronger understanding of their southerly neighbor (recall Mercer's derision). For example, Michael Scott of the Greater Yellowstone Coalition recalled that Y2Y had taught him "a lot about Canada and about how folks operate" there, while Willcox said that

Y2Y has made it easier for her to understand "what's happening on the Canadian side" and to "navigate my way around some of the information and the knowledgeable experts." Francis noted that Y2Y had enabled "acceptance of the different political realities" between U.S. and Canadian participants. "We weren't even talking to each other across the 49th parallel before Y2Y," she said, "and now we sit down and meet two or three times a year, we know what each other is doing, and we know the different political realities of each country."

In this regard, several participants described Y2Y as a complementary north-south arrangement, a theme I had heard repeated at several meetings. As Willcox put it, "We [in the United States] need the inspiration from the stories of the north. And they in turn need to learn the lessons of how we lost so much wild country to development, so they don't repeat that path. So it seemed obvious that there were some reciprocal relationships, which if developed, would result in mutual benefit." But at the same time, she emphasized, Y2Y has taught U.S. conservationists that "all is not secure" for wildlife in Canada—and that contrary to widespread belief, threats to wildlife in their northerly neighbor were rapidly increasing. As Ament has similarly noted, Y2Y has helped to dispel the "myth of abundance in Canada" and has helped alert U.S. conservationists "that they've got serious problems in southern Alberta and southern British Columbia."

The second type of transborder learning—"strategic learning"—alludes to the expanded universe of conservation strategies and tactics that conservationists were exposed to through participation in Y2Y. "The most important thing about Y2Y," Sawchuk said, was providing an opportunity "to see people up and down the Rockies and how they're doing their work." Many others expressed a similar point of view, and to a large degree this type of learning has been formalized through the Y2Y organization. In addition to the informal learning that takes place at CC/board and Council/partner meetings, Y2Y had by 2004 offered about 50 capacity-building workshops to over 250 conservationists representing about 65 network groups. These workshops focused on practical skills such as media skills and messaging, fundraising, organizational effectiveness, board management, and negotiations (Torgrimson 2004). It has also sponsored several more issue-oriented workshops, including one on "Managing Roads for Wildlife" and another on "Understanding Western Canada's Changing Economy."

Peacemaking in the Clan of Alpha

The Y2Y CC/board has been largely composed of leaders—often executive directors—of a variety of conservation organizations, and it was commonplace to hear these individuals described as alpha males, alpha females, or "alpha conservationists." Lifted from the scientific literature on the social relationships between wolves, these terms refer to the fact that each wolf pack has only one dominant pair of an alpha male and an alpha female. The challenge Y2Y faced from the start was not only to bring multiple alpha personas into the same den, but also to keep them there without violent consequences. "The environmental community is famous for being factious," as Ray Rasker pointed out, so this was no small challenge. Yet Rasker saw Y2Y as surprisingly successful in this regard: Under Y2Y, "a whole bunch of alpha males and females have met together three or four times a year for the last six or seven years, and they haven't killed each other." Similarly, but looking beyond the CC/board to the wider Council, Brian Peck noted that "getting 250 individuals or groups to work together toward a common goal in the environmental community is virtually unheard of." Yet again, Y2Y had succeeded in doing just that.

Coming together under Y2Y has thus helped to reduce tensions between different individuals and groups. Peck characterized Y2Y as leading to a "lessening of the divisiveness, lessening of hostilities among conservation groups, and a huge buildup in trust." Y2Y associate director Christine Torgrimson said that without Y2Y "there would definitely be more struggles over turf," and that Y2Y ameliorates those turf struggles that continue to arise. Ament recalled that before Y2Y there had been "flat-out competition" between environmental groups and individuals on certain issues—such as wilderness bills and approaches to endangered species protection—and that individual organizations had "been fighting with each other over the appropriate means to attain conservation goals." Y2Y had reduced some of those tensions, he said, by bringing the groups together to operate jointly on a common goal. "We had never said, 'Gee, how do we put a bigger picture together so that we all can achieve our own organization's goals yet still work together on mutually beneficial projects and campaigns?'" Overall, Ament summarized, "in a region that was known for its infighting," Y2Y has "brought a lot of us in contact with each other regularly, and that's been good for working relations among the groups."

Increased Funding for Conservation

Many conservationists have undoubtedly been attracted to Y2Y by the possibility of increased institutional funding. This should not obscure the idealism of the average Y2Y participant; highly talented and self-motivated to a person, any one of them could have been working in much more stable employment situations, taking home substantially larger paychecks. Yet immersed in a hectic and relentless whirlwind of lobbying, lawsuits, biological research, and comment period deadlines, they typically sustain their individual organizations on a precarious treadmill of discreet and relatively small grants. Consequently, a good deal of their time is spent worrying about where they will find next year's funding (and in several cases, this year's funding). So while they have not gone into conservation "for the money," finding the requisite financial support for their respective organizations constitutes a powerful undercurrent in how they go about their work—including the volunteer time they devote to Y2Y.

In the early years of Y2Y, no few conservationists had been concerned that Y2Y might actually divert funding from groups already operating in the region. Indeed, I heard it said several times that Y2Y originally elicited so much "interest" for that very reason—that some conservationists were participating in Y2Y to make sure that this "big idea" would not result in less funding for groups already operating in the region. But many conservationists seemed to have recognized early on that Y2Y constituted a potential opportunity to find new funding sources. Or as Sawyer skeptically argued, in Y2Y's early years, many conservationists came with the expectation that "Y2Y was going to just sort of rain money down from above to help all these local groups with cash. So a lot of people's involvement in Y2Y was not entirely because they were committed to the concept or entirely that they were altruistic. It was that they were saying, 'Well, if we're involved and we're at the core of the action here, we will get the benefit.'" Accordingly, Y2Y staffer Robinson noted that participation in Y2Y meetings spiked whenever new grant-making possibilities were announced. In addition, a few conservationists privately (and revealingly) complained that all their volunteer work on Y2Y had led to little financial benefit for their respective organizations.

Regardless of the inevitable cynicism concomitant with the subject of money, these concerns do raise an important question: How much additional conservation funding has Y2Y brought to the region? Gary Tabor of the Wilburforce Foundation has estimated that since Y2Y's conception, about twenty foundations have put a minimum of $45 million into the region as a result of their attraction to and belief in the Y2Y vision. Rasker described this phenomenon as an "implosion" that drew in numerous funders that had previously not been interested in the Northern Rockies. With funding from the LaSalle Adams Foundation, the Y2Y organization itself set up a "Y2Y minigrants" program for grassroots activists working in the Y2Y region. Awarding more than 120 grants for a total of over US$180,000, Y2Y literature states that "these small grants pack a big punch, allowing Y2Y partners to conduct scientific research that can impact public policy and build much-needed support among key audiences" (Torgrimson 2004, 6). Overall, these sums help to substantiate Bob Ekey's conclusion that Y2Y has acted as "a rising tide that's lifted all boats for funding."

In explaining its fundraising power, Locke characterized Y2Y as a magnet of mutual attraction between funders and conservationists. For foundations that are perpetually "besieged by good opportunities," Y2Y offered a "big picture" that constituted a better way to organize their funding strategies. Funders are attracted to Y2Y, he said, because they "hate to make winners and losers—and here's something everybody wins at." On the conservationists' side, Locke pointed out that "if activists can't find funding for an idea, they move to a different idea"—and the very fact that conservationists have not "moved on" illustrates the persistent fundraising power of Y2Y. Several others pointed to this symbiotic effect between grant makers and grantees. Craig Stewart, for instance, argued that Y2Y has "given a lot back to the funders because you've got people that are doing things in concert now," while Robinson noted that foundations approved of Y2Y's "operating principle" of enabling conservationists to do something together "that we wouldn't or couldn't do by ourselves."

Y2Y has also been an important tool for educating funders. Mike Bader of the Alliance for the Wild Rockies, for instance, described Y2Y as playing a key role in "informing many people in the foundation world, the media, and the private land trusts of the concepts of conservation biology and the need to protect corridors and some private lands." Legault pointed out that "one of the great things that Y2Y's done is to provide us with the context to go to funders which otherwise would not give a damn about some little area in west Alberta." With Y2Y, he continued, funders understand that "by protecting this valley, we're freeing up wildlife movement for a much larger region." Ament made a similar point, noting that Y2Y had helped American Wildlands build on the fame and reputation of

Yellowstone National Park—even though the organization did not focus on that park's particular issues. "It's those lands in between the famous places that are the hardest thing to raise money for," he said. "Just talk to a group that's trying to protect the Big Belts or the Tendoys; you don't go into Miami and say, 'You've got to save the Tendoys,' because they don't know what the hell you're talking about." Y2Y, however, had in a way made those areas relevant to the long-term fate of Yellowstone, thereby giving conservationists "a great way to say all of these things are components."

Science and Conservation Planning

As described previously, the science of conservation biology was a principal influence in Y2Y's "web of conception." Yet even as conservation biology has been "fundamental to conservation in the Yellowstone to Yukon region" (Herrero 1998), Y2Y has, in turn, spawned and inspired a tremendous amount of scientific work throughout the region. And it did so from the start: Coming out of the first Y2Y meeting in 1993, Professor Tim Clark of the Yale School of Forestry and the Northern Rockies Conservation Cooperative recruited attendees (and others) to contribute to a special issue of the journal *Conservation Biology* on "large carnivore conservation in the Rocky Mountains of the United States and Canada." By April 1996, conservationists and scientists had begun actively discussing the development of a "rapid assessment" to inventory the resources of Y2Y (Y2Y Biodiversity Strategy 1996a). Originally described as a "kind of a state of Y2Y" (Y2Y Coordinating Committee 1996) and considered a first step toward the goal of creating a broad-scale "conservation plan" for the Y2Y region, that project subsequently transmogrified into a biological "atlas" for the region. A rough draft of the atlas was available at the formative 1997 "Connections" conference (Y2Y Conservation Initiative 1997a), and the final version was released the following year under the title *A Sense of Place* (Harvey 1998). The 138-page atlas included the contributions of more than twenty-five scientists and conservationists and covered subjects that ranged from a broad description of the region to an overview of the region's ecological and economic status.

By 1997, Y2Y literature clearly stated that one of Y2Y's principle goals was to "enable, inspire and energize" conservationists to use the "principles and practices of conservation biology" throughout the Y2Y region (Robinson 1997). And although a few voices questioned the need for more scientific research (more than once I heard some variant of "we already know what we need to do to save the grizzly bear"), most Y2Y participants have promoted or accepted the importance of science to conservation. Yet despite such consistent support, Y2Y-related science has progressed along a circuitous route—a result, at least in part, of the divergent opinions on where and how to focus Y2Y-supported research. One of the principal points of contention was whether to focus solely on the broader region as a whole (and indeed, whether such a broad focus was even possible) or to focus on particular locations, species, and threats. This particular tension was resolved early on—somewhat—with a decision to keep the perspective broad but also to divide the Y2Y region into seven subregions, each of which was to develop a detailed conservation map portraying the necessary protected core areas, corridors, and transition zones for wildlife protection.

More problematic were the unrealistic assessments of how much work Y2Y's science agenda would take, as indicated by an early wishful title of "Conservation Plans 2000" (Y2Y Conservation Initiative 1998, Y2Y Coordinating Committee 1999). It also became clear that conducting seven subregional plans was simply too expensive. Furthermore, the idea of creating "a single definitive map" as a broad conservation plan turned out to be unrealistic due to the drastic disparity in quantity and quality of scientific data across the region. Nevertheless, there were some feasible investigations that focused at the "Y2Y scale." So with the creation of a formal Y2Y Science Program in 1999, a "parallel tracks" approach was adopted to support investigations looking both at the Y2Y region as a whole and at the subregional scale. A third track was subsequently added, focusing on how to apply all the resulting scientific information to specific conservation campaigns. Each of the three tracks was then applied to four conservation "processes": (1) the Multi-Carnivore Conservation Process, (2) the Grizzly Bear Conservation Process, (3) the Bird Conservation Process, and (4) the Watershed Conservation Process (Bienen and Mahr 2003). As this strategy was being worked out, Y2Y partnered with the Wilburforce Foundation in 1999 to establish the Yellowstone to Yukon Conservation Science Grants program, which over the following five years would fund 46 scientific research endeavors throughout the region (Mahr and Mauro 2004, 3).

No aspect of Y2Y has changed as consistently or as rapidly as its science program, and it is likely that the program's format will have further changed by the time this book makes it into print. At this point, then, it is worth taking a step back to review a more theoretical assessment of the relationship between Y2Y and conservation science. Before 1997, Y2Y had begun to take on the mantle of what Peter Haas has called an "epistemic community." A widely used term in the literature on international environmental diplomacy—particularly as it relates to how international environmental agreements come about—an *epistemic community* has been defined as "a network of professionals with recognized expertise and competence in a particular domain and an authoritative claim to policy-relevant knowledge within that domain" (Haas 1992, 3). Although epistemic communities have taken credit for international action on environmental issues (see Bernstein 2001 for a balanced perspective on this question), Suzanne Levesque (2000, 279) in an extensive treatment of Y2Y science argued that it was the use of scientific knowledge by a wide range of actors—and not a "community of scientists" per se—that both defined and gave power to Y2Y:

> It is environmental advocates, and not environmental scientists, that have primarily guided the network's development from its earliest days. Although an epistemic community of biodiversity scientists was involved in the network at its inception and in its infancy and played a large role in defining the problems of biodiversity preservation in the region and in formulating the Y2Y proposal, the scientific community's direct, intense involvement in the network's decision-making processes was short-lived. Considerable differences in personalities, worldviews, linguistic formulations, working styles and temporal scales exist between and among network scientists and advocates and these factors continue to render intensive, long-term interaction challenging. However, the need for scientists to maintain a "firewall" between themselves and environmental advocates in order to avoid having their legitimacy and credibility challenged by other scientists, governmental and academic institutions and the public appears to represent the primary reason many scientists are not more directly involved in network decision-making.

Written in 2000, just after the peak of excoriation of Y2Y by prodevelopment forces (Chester 2003b), Levesque's critique was on the mark (see also Levesque 2001). However, while it remains true that many biologists and ecologists working in the Y2Y region have not associated themselves with Y2Y, a small number of scientists have continued to play an "intense" role in framing its scientific agenda. Further, Y2Y-sponsored scientific conferences and workshops attract a broad range of scientists working throughout the region (Mahr and Mauro 2004). Describing these sessions as "Y2Y science summits," Locke argued that Y2Y was not only gathering together "the who's who" of conservation biology but had also provided the region's first practical mechanism for coordination between aquatic scientists, grizzly bear biologists, and ornithologists.

Many other Y2Y participants also pointed to the important learning and networking functions that Y2Y has served by providing a tangible forum for connecting the scientific community with conservation advocates. Y2Y's first two employees, Robinson and Aengst, both argued that it was a strategically fortunate connection; without Y2Y, they said, conservationists would have been far less likely to know about—much less incorporate—powerful science-based arguments in their conservation advocacy. Robinson also noted that an increasing respect for Y2Y's role in networking scientists has led more mainstream environmental groups—groups previously reluctant to be associated with Y2Y—to engage publicly with both scientists and conservationists working within the Y2Y network. He added that although many land trusts are still wary of working with Y2Y, some have begun to use Y2Y-sponsored scientific analyses of the region to help them identify high-priority areas for conservation easements and acquisitions. Lewis described these particular land trusts as being "more open to coordinating their efforts, to thinking about their work in a larger landscape context, and to tying into Y2Y information and mapping." In addition to land trusts, at least one prominent private landowner has directly responded to priority areas defined by Y2Y science (see, for example, Phillips 2000, 2001).

In sum, as Levesque points out, Y2Y could be characterized as a symbiotic tie between scientists and activists, between scientists and the public, and between scientists and the government. In the first case, scientific credibility is necessary to the conservation community's desire to be taken as a competent "player," while scientists need activists to "achieve goals they cannot promote independently without sacrificing social and political legitimacy and credibility" (Levesque 2000, 282–283). In the second case, Y2Y participants act as "knowledge brokers" between the scientific community and the public, and in so doing "reframe complex scientific and social issues and parochial concerns, interests and identities simply, symbolically, evocatively and in terms of universalistic values" (281). In the final case, scientific knowledge fostered by Y2Y has allowed scientists and activists to engage with government agencies (and courts) in order to have an impact on "decision processes by holding environmental policy-makers accountable for actual or proposed decisions under various national, state or provincial laws, or policies" (280).

Regional Examples: The Muskwa-Kechika and the Yukon

When I asked Y2Y participants whether specific subregions had benefited from the Y2Y vision, many brought up the Muskwa-Kechika and/or the Yukon as places where Y2Y had achieved conservation victories. Yet a close look at these two cases reveals the difficulty in making causal connections between idea and action.

As noted previously, Canadians understand the phrase "Northern Rockies" to signify the vast wilderness area in north-central and northeastern British Columbia where the province's Ministry of Forests had identified over fifty adjoining watersheds that were "virtually pristine," each with an area greater than 50 square kilometers (just over 19 square miles) (Sawchuk 2004, 18). The area has come to be known as the Muskwa-Kechika, these being the names of two large rivers flowing out of the region, and has been described both as the "largest intact predator-prey system in North America" and as "one of the few remaining large, intact, and almost unroaded wilderness areas south of the 60th parallel" (Gailus 2000, 19; Mitchell-Banks 2003, 104).

Although a few conservationists and local guides had called for protection of the region over the past decades, it was not until 1992 that Wayne Sawchuk of the Chetwynd Environmental Society (CES) and George Smith of CPAWS launched the "Northern Rockies—Totally Wild" campaign to protect the Muskwa-Kechika. It was a complex and atypical campaign that quickly became tied to the negotiations for three regional Land and Resource Management Plans (LRMPs), a localized planning process that the B.C. government had established to allow for local stakeholder groups to work out long-term management plans among themselves. In addition to conservationists, LRMP stakeholder groups included the forest industry, the oil and gas industry, snowmobilers, hunters and anglers, and labor and community representatives. Due to concern over future negotiations regarding aboriginal rights to the land—Canada's system of treaty-making with First Nations is notoriously complex (see Dickason 1997)—First Nations participated only as observers (Sawchuk 2004).

Several years of negotiations led to the agreements over the Fort Nelson LRMP and the Fort St. John LRMP in 1997, and the agreement over the Mackenzie LRMP in 2000 (Mitchell-Banks 2003, 104). In turn, contiguous sections of each of these LRMPs were carved out to form the Muskwa-Kechika Management Area (MKMA), which encompassed a total area of 64,000 square kilometers (24,324 square miles). About 25 percent of this area was protected under 20 provincial parks with the remaining set off as "special management zones" in which, with limited exceptions, resource extraction was to be highly controlled and remediated (Muskwa-Kechika Information Office 2005a, 2005b; Mitchell-Banks 2003). According to CPAWS (2001), the various stakeholders had "helped achieve a conservation achievement unprecedented in North America— a consensus-based, multi-stakeholder agreement to establish the largest conservation system on the continent, reflecting the scientific tenets of conservation biology." Although provincial government support for maintaining the standards set under the original agreements has wavered since 2000, the MKMA arguably set a geographic precedent for conservation unparalleled in the North American continent.

In both formal presentations and informal conversations, Y2Y received credit for playing a significant role in the establishment of the MKMA. The claim has also appeared in print: Wilkinson (2000), for example, quoted a forest industry representative regarding his positive assessment of Y2Y:

> Mike Low, general manager of Abitibi Consolidated Inc., a large forest products company in British Columbia, says that industry was caught up in the spirit of Muskwa-Kechika being an important piece in the Yellowstone to Yukon puzzle which is trying to chart a new approach to conservation along the Rocky Mountain front. Low, who admits to being initially skeptical, views the Mackenzie agreement as a breakthrough that respects his desire to earn a living from the forest and leave a legacy of wildness for his grandchildren.

Wuerthner (2001a, 15) also credited Y2Y for having "affected conservation efforts along the spine of the continent in Canada," particularly the "Y2Y-inspired protective management regime" in the Muskwa-Kechika region. And Locke noted that at some time before the decision, Brian Churchill, the regional head of the B.C. government's wildlife division and member of the Fort Nelson LRMP, had heard independently about Y2Y and had been inspired to push the Y2Y approach in the land-use planning process. Locke felt strongly enough about Churchill's role to state that without Y2Y, the land-use decision probably would not have happened. "Nobody reached Brian on a one-on-one conversation," Locke said. "He got there because he heard about it, it made sense to him, and that was what he wanted to do." Churchill affirmed Locke's assessment, adding that whereas the idea of "one big park had no traction in the region, the concepts of wildlife corridors and connectivity embodied in the Y2Y vision had a lot of traction."

However, many other observers of the process were far less willing to attribute such a causal relationship from Y2Y to the MKMA decision. For example, B.C. conservationist Anne Levesque flatly stated that credit for the Muskwa-Kechika lies with the "amount of sweat equity of the local people" and not with the Y2Y vision. Ray Rasker of the Sonoran Institute credited the work of Sawchuk and Smith who "worked for years with the First Nations, they worked with the oil and gas people, they worked with mining, they worked with timber." It was at a *local* level, said Rasker, that they formed "a broad coalition that you could not defeat." Most skeptical on this point was Mike Sawyer, who argued that although the Muskwa-Kechika campaign "was sold in the context of its location in the Y2Y," any assessment of the causal relationship would be "pure speculation."

Most Y2Y participants took neither of these positions, instead describing Y2Y and the MKMA decision as having either a complementary or a reciprocal relationship. For example, Jeff Gailus, a former Y2Y employee who has looked extensively at the Muskwa-Kechika LRMPs, attributed both Y2Y and the MKMA decision to the same principles of conservation biology, noting that "it's no surprise the MKMA is being compared to the Yellowstone to Yukon Conservation Initiative" (Gailus 2000, 25). Others believe that Y2Y had provided Smith and Sawchuk with a highly networked community of enthusiastic supporters—a community that, as Legault put it, inspired them "to do battle for five long hard years in the Northern Rockies." Barb Cestero of the Sonoran Institute argued that Y2Y helped to attract outside attention to the Muskwa-Kechika while at the same time educating Y2Y participants—particularly those in the United States—about this vast natural area.

As both a principal actor in the Muskwa-Kechika negotiations as well as a consistent participant in Y2Y CC/board meetings, Sawchuk largely corroborated this middle perspective. "The Muskwa-Kechika is an important part of the Y2Y vision," he said, and "certainly we had that in mind and we were working on it." The Y2Y network had supported the Muskwa-Kechika campaign "at key points in particular," he said, and gave the Muskwa-Kechika campaign positive publicity, including "exposure to the funding agencies." Furthermore, the "Muskwa-Kechika set an example of what could be accomplished, an example of how to implement a conservation vision on the ground."

Yet Sawchuk also emphasized that his involvement with Y2Y principally stemmed from it being "the right thing to do"—and not from it being essential to the LRMP processes. Indeed, he went so far as to state that his work might have been less onerous without Y2Y:

> On a local level, there's a fair amount of, not antagonism, but ill-feeling towards Y2Y generated by the forest industry. So we have to deal with that, and it's basically causing us some trouble. So our project was going along just fine without Y2Y—we don't invoke Y2Y, we were doing our work for the most part certainly on a local level. And so I think probably without Y2Y it would be—if anything—easier.

Ultimately, as Sawchuk summed it up, Y2Y "didn't turn the tide or make a critical difference, but it was additive to what we were doing."

Whereas the Muskwa-Kechika has historically been hidden even to Canadians, citizens of both Canada and the United States find the Yukon on their cultural radar screens. The average Yukoner, in turn, would like to keep the rest of the world on the Yukon's own radar screen—meaning, that is, "outside" of the Yukon. Prominent Canadian conservationist Juri Peepre, a former executive director of CPAWS-Yukon, has noted that common usage of the word *outside* in the Yukon (as in going "outside" to visit relatives down south in Edmunton) reflects a widely held distrust by Yukoners of anything from the "outside"—a distrust that can be hard to overcome, particularly for conservationists who believe in working at a transborder landscape level. On the other hand, noting that many Yukoners have traveled to Banff and Jasper National Parks at some point in their lives, Peepre pointed out that they are familiar with the Rocky Mountains and that consequently Y2Y is a "familiar ecosystem." Peepre further described the utility of Y2Y for the Yukon in terms of the territory's love-hate relationship with the "outside":

> We often follow what the rest of the country is doing. We're in a remote region, and there's a double psychology here. On the one

hand, people resent the outside. Yukoners refer to "outsiders"—sort of like in the Maritimes where you're "from away"—it's the same psychology. So on the one hand, people resist being connected to places like the United States and Yellowstone. But on the other hand, the reality is that we mimic what other people do as well. It was our view that by being part of this big, international idea, we leverage support for the northern conservation agenda. As it happens, it was perfectly in line with the conclusions that we had already drawn here in terms of big wilderness with connected corridors and all these kinds of things. It was a gift to have this idea so well connected to what we were thinking.

To counter the Yukon's inward-looking perspective, Peepre said, CPAWS-Yukon's goal has been "to bring in the global connections all the time and say, 'we're connected to the rest of the world.'" Looking back to the first meeting of Y2Y in 1993, when the first iteration of the Y2Y map showed the area ending just over the Yukon's southern border with B.C. (Locke 1994), Peepre recalled that he went to that meeting to argue that the area should be expanded to include much more of the Yukon, as well as part of the Northwest Territories—a proposal that was immediately accepted. For the Yukon, according to Peepre, there were:

> great advantages for the whole Y2Y concept in general based on the premise that in the north here we had a reservoir of truly pristine ecosystems that could serve as a benchmark for the whole rest of the system. And that was one of the reasons we were very interested in it. But we also saw that in order to protect the North, by being connected to places like Yellowstone and the Rocky Mountains, we could learn from those experiences.

Elaborating on this point, activist and educator Larry Gray, a former staffer with CPAWS-Yukon, characterized Y2Y as very helpful in bringing this message home by contrasting the relatively unscathed Yukon with the problems of its southern neighbors (including British Columbia and Alberta with the United States). Using the example of oil and gas wells and associated seismic lines, Gray said that through Y2Y we can now warn, "look what's happened down there; this is going to happen in the Yukon if we don't do things right." Overall, Gray summarized, an important achievement of Y2Y has been "just getting this whole thing kind of in the public consciousness" in the Yukon. Accordingly, Y2Y has been featured in a number of conservation publications in the Yukon, including one listing Y2Y as one of eight "priorities for 2000" (CPAWS-Yukon n.d.; see also Bennett et al. 1999; Locke 2000, 112–113; Yukon Wildlands Project 1996–97).

While Peepre concurred with this general line of reasoning, he also pointed out that "Y2Y itself has not been a driver" in terms of the day-to-day work of Yukon conservationists. Although Peepre had "no doubt" that Y2Y was tied directly into conservation campaigns in the south, he observed that "it's not that way here." The Y2Y vision has been "supportive and inspirational and integrated," but conservationists in the Yukon have not "led our campaigning with Y2Y because we could not afford the luxury of participating daily in this big vision when we had ground campaigns to win."

Nevertheless, Peepre did describe the Y2Y vision as "a unifying theme" in the Yukon beyond the conservation community. Y2Y has "been a way to bring communities together to think about the broader landscape," he said, "and it has influenced the way people think about protected area boundaries and the scale of things that we need." As a concrete example of a Y2Y-inspired connection, Peepre pointed to the critical role of Y2Y in achieving "momentum on the transboundary issues with Northwest Territories." In 1998, CPAWS-Yukon, CPAWS–Northwest Territories, and the Liidlii Kue First Nation government of the larger Deh Cho First Nations government held a workshop entitled "A Northern Vision for the Yellowstone to Yukon Conservation Initiative" for First Nations communities throughout the "Y2Y north region" (Liidlii Kue First Nation et al. 1998). Peepre ascribed the high level of interest in the workshop to people's curiosity over Y2Y, and noted that there has been significant follow-up work to the workshop. He also made it clear that the workshop not only strengthened relationships between conservationists and First Nations, but also helped to build up the conservation community within the First Nations themselves. Less than a year later, Ray Rasker of the Sonoran Institute came to the Yukon to do give a public talk and an economics workshop on sustainable communities (Rasker 1999), which according to Peepre "gained tremendous credibility" for Y2Y even among government officials. Peepre described Rasker's involvement as a "direct benefit" of Y2Y since the two had only come into contact through Y2Y meetings.

In 2002, Peepre was also involved in another significant development of Y2Y's outreach to First Nations when representatives from three aboriginal groups attended a "Y2Y-north workshop" on regional land-use issues (Y2Y Conservation Initiative 2003b). The traditional territory of these groups, specifically the Deh Cho, Kaska Dene, and Tetl'it Gwich'in First Nations, included lands in the Yukon, northern B.C., and the Northwest Territories (CPAWS 2002). The "highlight" of the meeting was a letter of understanding signed between the Kaska Nation, CPAWS-Yukon, and CPAWS-British Columbia that called for cooperative efforts "to develop an ecologically sustainable economy and to designate a new national park in the Kaska Nation's globally significant traditional territory." The letter also explicitly recognized the Kaska Nation's support for the Y2Y "in the context of protecting the ecological and cultural integrity of their traditional territory."

Y2Y Today and Tomorrow

As propounded in the tipping point theory, the concept of "social epidemics" resembles that of *memes*, which have been defined as ideas "that become commonly shared through social transmission" (Aunger 2000, 2). In proposing memes, biologist Richard Dawkins (1976, 206) wrote: "If a scientist hears, or reads about, a good idea, he passes it on to his colleagues and students. He mentions it in his articles and lectures. If the idea catches on, it can be said to propagate itself, spreading from brain to brain." This seems to be exactly what is happening with Y2Y—but hardly confined to the halls of academia. Y2Y has been the primary or secondary subject in hundreds of print publications and other media sources, and reference to it in countless lectures and conversations continues to rise with no apparent sign of diminishment.

At a bare minimum, the Y2Y meme has successfully established itself within the conservation community, and shows very strong signs of propagating itself beyond that often cloistered community into the indefinite future. Although it is difficult to measure what constitutes a "very successful meme," it is fair to say that the Y2Y meme would be very successful if it became as identifiable as, say, *the* Serengeti or *the* Great Barrier Reef. Both of the latter are real places, but they are also iconic ideas that are extensively transmitted in Western culture—the first being a vast African trove of large charismatic animals, the second an underwater wonderland of color and infinite shapes. In other words, they are successfully replicating memes—memes with the advantage of a much longer history than that of Y2Y. Yet it is reasonable to predict that if Y2Y stays on the trajectory it has been on for the last decade, it could soon become equally well recognized. That would constitute the Y2Y vision's first giant step toward becoming the type of *story* that creates "meaning, values, and a human order that sees beyond the short term" (Johns 2003a).

As we have seen, the robustness of the Y2Y meme derives partly from its composition of two longstanding icons: Yellowstone as a symbol of American heritage, the Yukon as a symbol of Canadian heartiness. Nevertheless, Y2Y yet remains a fledgling conglomeration, and what it has accomplished despite its recency is revealing of its power: In less than a decade of public exposure, Y2Y has built a foundation for itself as a commonly understood entity. It has, in language that many Y2Y participants would consider crass, "branded" itself. Yet Y2Y participants generally understand how far Y2Y has come, and a commonly heard refrain is that "we could all be hit by a bus and Y2Y would live on."

That may be the case, but it may equally be a case of overconfidence. The danger of such overconfidence is that it could condone a passive approach to promoting the Y2Y vision. Consequently, Y2Y participants might be well advised to adhere to the advertising experts, who warn that branding is an interminable task; why, otherwise, would we have to endure the incessant battling between Coke and Pepsi? Again, since branding implies a superficial mark, whether on the hindquarters of a cow or a TV watcher's brain, some will chafe at applying the word to Y2Y. But that is largely a semantic point. Whether you call it branding or giving form to core values, the task of maintaining widespread recognition of Y2Y will, at least for quite some time, rely on its avid proponents. For if the Y2Y bus were to crash, if the directed effort to promote Y2Y were suddenly to stop, there is absolutely no guarantee against Y2Y being dumped into the junkyard of yesterday's big ideas. Fortunately, there is little indication that the Y2Y bus is about to drive off the road: Its central organization is strong and relatively well funded, many conservationists throughout the region are participating in Y2Y activities, and Y2Y continues to be publicized in the media (for an extensive treatment of Y2Y's own assessment of its effectiveness, see Torgrimson 2004).

In one of the earlier analytical treatments of Y2Y, Clark and Gaillard (2001, 237) recommended that "Y2Y members embark on a process to build a shared problem orientation, capitalize on diversity, and employ adaptive management explicitly and systematically." While these recommendations remain highly relevant, another way to approach a phenomenon as complex as Y2Y is to point out the most pressing questions facing its future.

The first question is operational: How does Y2Y make progress on its mission while working with so many different and varied members, and can it be open to new participants from outside the conservation world without compromising the principles of its formative participants? As Tabor (1996, 8) recognized early on, "taking the concept beyond the environmental community will be the true challenge of Y2Y."

The second question follows from the first: Even if Y2Y chooses to remain institutionally within the conservationist camp, how can it ensure democratic participation? Several Y2Y participants reflected Skeele's strong concern over Y2Y becoming distanced from grassroots conservation advocates: "We're supposed to have a big tent, we're supposed to have lots of people represent us—and I wonder whether we're really doing that or not. And that is a concern of how empowering [Y2Y] will be. That gets right back to the initial concern that it's just going to be a number of people creating this tent but really dictating the situation—and that isn't really all that democratic and supportive and encouraging. That's my biggest concern for Y2Y." Meetings of the CC that I attended reflected these challenges of diversity. Procedural issues are discussed and rediscussed, substantive differences are debated and then debated again, and frustration often rears its ugly head. The relatively recent changeover from a "Coordinating Committee" to a formal board has modified the dynamics of these frustrations, but it has not terminated them.

The third question goes beyond issues of institutional structure: How can Y2Y best counter perceptions that conservation spells economic doom for the region? In light of ominous demographic and societal trends, finding the answer to this question may be the most difficult task of the Y2Y effort. Notably, while Y2Y will have to focus externally—that is, on society in general—the real challenge will be internal due to a broad philosophical split between Y2Y participants. Some argue that a central task of Y2Y is to ensure that the extractive sectors of the region's economy—in particular mining, forestry, and grazing—begin to operate in a sustainable manner throughout the region. They argue that Y2Y should attempt to reach out to members of those extractive sectors of society, particularly since many of the individuals involved in them live in the region for the express purpose of taking advantage of its environmental amenities. Others argue that Y2Y needs to concentrate on the original and fundamental goal of protecting habitat across large landscapes. Creating sustainable communities, they believe, is important but secondary to Y2Y's mission. To be fair, the strict dichotomy between these two positions has significantly softened over the past decade. Yet the fundamental question remains: Can Y2Y bring these two worldviews together so as to offer a unified and even more powerful conceptualization of what the region might become?

The fourth and last question is grand and encompassing: What should Y2Y make of its own precedent? It is the question Y2Y participants grapple with whenever they attend Y2Y meetings, whenever they weigh in on subcommittee conference calls, whenever they think about Y2Y. It rests on the premise that Y2Y has indeed set a precedent as a specific landscape-scale conservation initiative. And on that point, few would disagree—even those skeptical of Y2Y's on-the-ground accomplishments, such as noted grizzly bear biologist David Mattson, who allowed that the precedent of Y2Y has "scaled up conservation issues to a level that is critically important." Consequently, other regions have picked up not only Y2Y's large-landscape approach, but also its alphanumeric acronym; these include the Algonquin to Adirondacks Conservation Initiative (A2A) (CPAWS 1998, 2004); the Baja to Bering Sea Marine Reserve Initiative (B2B) (Center for Marine Conservation 2000; Jessen and Lerch 1999); the Bhutan Biological Conservation Complex (B2C2) (Sherpa et al. 2004); the Highlands to Ocean Initiative (H2O) (Hiss and Meier 2004); the Mount Agamenticus to the Sea Conservation Initiative (MtA2C) (Schmitt 2005); and the Quabbin to Cardigan Collaborative (Q2C) (Stephens 2004). Y2Y has, it seems, set a precedent in more ways than one.

The "grand experiment" of Y2Y, as one funder put it, is thus leading the way in landscape conservation. While that is a point of pride, it must also be a point of concern. For if Y2Y fails, if the region's protected areas are not ultimately protected, if the communities of the region compete to outgrow each other, if human land use outstrips the needs of wildlife, then what hope for the continent, for the world, will emanate from this grand experiment?

References

Alliance for the Wild Rockies. 1995. http://www.wildrockiesalliance.org/ActivOrg/AWR/AWR.html. Last accessed November 1997.

Alliance for the Wild Rockies. 2001. AWR Member Groups. http://www.wildrockiesalliance.org/about/membergroups.html.

American Wildlands. n.d. Corridors of life science report. Bozeman, MT: American Wildlands.

Aunger, Robert. 2000. Introduction. In *Darwinizing culture: The status of memetics as a science*, ed. Robert Aunger, 1–23. Oxford, England: Oxford University Press.

Bader, Mike. 1992. A Northern Rockies proposal for Congress. Special issue, *Wild Earth*: 61–64.

Bader, Mike. 1998–99. NREPA: Ecology meets politics in the Northern Rockies. *Wild Earth* 8, no. 4: 78–80.

Barcott, Bruce. 2001. For God so loved the world. *Outside Magazine* (March).

Bengeyfield, Pete. 2003. *Incredible vision: The wildlands of Greater Yellowstone*. Helena, MT: Riverbend Publishing.

Bennett, Graham. 2004. *Integrating biodiversity and sustainable use: Lessons learned from ecological networks*. Gland, Switzerland: World Conservation Union (IUCN).

Bennett, Bruce, Dave Mossop, Rhonda Rosie, Alejandro Frid, Randi Mulder, Brian Slough, Dave Jones, Juri Peepre, and Marty Strachan. 1999. *Wolf Lake area: A preliminary report on the findings of 3 biological surveys at Nisutlin Lake, Wolf River and Morris Lake*. Wolf Lake Report No. 2, CPAWS-Yukon Research Report No. 5. Whitehorse, Yukon: Yukon Chapter of the Canadian Parks and Wilderness Society.

Bernstein, Steven F. 2001. *The compromise of liberal environmentalism*. New York: Columbia University Press.

Bienen, Leslie, and Marcy Mahr. 2003. *The first four years: A brief history of Yellowstone to Yukon's Conservation Science Program*. Files of the Yellowstone to Yukon Conservation Initiative, Canmore, Alberta.

Botkin, Daniel B. 1995. *Our natural history: The lessons of Lewis and Clark*. New York: Putnam.

Boyd, David R. 1993. Minutes, Wildlands Project/Yellowstone to Yukon Biodiversity Strategy Meeting, Kananaskis Field Station, Alberta. Files of the Yellowstone to Yukon Conservation Initiative, Canmore, Alberta.

Burke, David. 1997. Trekking the continent to spread the Y to Y message. *Canmore Leader*, October 7, A6.

Center for Marine Conservation. 2000. Marine reserves from Baja to the Bering Sea. *Marine Conservation News* (Summer): 4.

Chadwick, Douglas H. 2000. *Yellowstone to Yukon*. Washington, DC: National Geographic Society.

Chester, Charles C. 2001. Review of *Yellowstone to Yukon* by Douglas Chadwick. *Conservation Perspectives: The Online Journal of the Massachusetts Chapter of the Society for Conservation Biology*. (Winter). http://www.MassSCB.org/epublications/winter2001/chester.html.

Chester, Charles C. 2003a. Creating an international community through the Internet: The Yellowstone to Yukon Conservation Initative. *Report on Conservation Innovation.* (Winter). http://harvardforest.fas.harvard.edu/research/pci/Publications.htm.

Chester, Charles C. 2003b. Responding to the idea of transboundary conservation: An overview of public reaction to the Yellowstone to Yukon (Y2Y) Conservation Initiative. In *Transboundary protected areas: The viability of regional conservation strategies,* ed. Urami Manage Goodale, Marc J. Stern, Cheryl Margoluis, Ashley G. Lanfer, and Matthew Fladeland, 103–125. New York: Food Products Press (published simultaneously in the Journal of Sustainable Forestry, 17: 1–2).

Clark, Ella Elizabeth. 1966. *Indian legends from the Northern Rockies.* Norman: University of Oklahoma Press.

Clark, Tim W., and David L. Gaillard. 2001. Organizing an effective partnership for the Yellowstone to Yukon Conservation Initiative. *Yale School of Forestry & Environmental Studies Bulletin* No. 105. http://www.yale.edu/environment/publications/bulletin/105pdfs/105clarkgaillard.pdf.

CPAWS. 1998. Algonquin to the Adirondacks: Following the path of the lynx. Canadian Parks and Wilderness Society. *The Wilderness Activist* (Spring), 7.

CPAWS. 2001. *Background on the Northern Rockies/Muskwa-Kechika.* Canadian Parks and Wilderness Society. http://www.cpaws.org/northernrockies/background.html. Last accessed June 10, 2005.

CPAWS. 2002. *Traditional territory of the Kaska Nation in British Columbia, Yukon, and Northwest Territories.* British Columbia: Canadian Parks and Wilderness Society, British Columbia Chapter. http://www.cpawsbc.org/boreal/map.php.

CPAWS. 2004. *Algonquin to Adirondack.* Ottawa: Canadian Parks and Wilderness Society, Ottawa Valley Chapter. http://www.cpaws-ov.org/A2A.htm.

CPAWS-Yukon. n.d. *Yukon wild!* Whitehorse, Yukon: Canadian Parks and Wilderness Society, Yukon Chapter.

Craighead, Frank C. 1979. *Track of the grizzly.* San Francisco: Sierra Club Books.

Craighead, Lance. 1998. Plot involvement comes as surprise to Web page author. *Bozeman Chronicle,* July 19.

Dawkins, Richard. 1976. *The selfish gene.* Oxford, England: Oxford University Press.

Dickason, Olive Patricia. 1997. *Canada's First Nations: A history of founding peoples from earliest times.* New York: Oxford University Press.

Ellis, Cathy. 1997. Conservationists to take grizzlies' trail. *Calgary Herald,* December 15.

Explore Magazine. 1998. Walking the talk. May-June, 12.

Finkel, Michael. 1999. From Yellowstone to Yukon. *Audubon* (July-August), 44–53.

Foreman, David. 1992. Developing a regional wilderness recovery plan. Special issue, *Wild Earth:* 26–29.

Foreman, Dave. 2004. *Rewilding North America: A vision for conservation in the 21st century.* Washington, DC: Island Press.

Gadd, Ben. 1995. *Handbook of the Canadian Rockies.* Jasper, Alberta: Corax Press.

Gadd, Ben. 1998. The Yellowstone to Yukon landscape. In *A sense of place: Issues, attitudes and resources in the Yellowstone to Yukon ecoregion,* ed. Ann Harvey, 9–18. Canmore, Alberta: Yellowstone to Yukon Conservation Initiative.

Gailus, Jeff. 2000. *Bringing conservation home: Caring for land, economies and communities in western Canada*. Bozeman, MT: Sonoran Institute and Yellowstone to Yukon Conservation Initiative. http://www.y2y.net/overview/Y2Y_Sonoran.pdf.

Gladwell, Malcolm. 2000. *The tipping point: How little things can make a big difference*. Boston: Little, Brown.

Gluek, Alvin Charles. 1965. *Minnesota and the manifest destiny of the Canadian northwest: A study in Canadian-American relations*. Toronto, Canada: University of Toronto Press.

Greater Ecosystem Alliance [Northwest Ecosystem Alliance]. 1994. The Wildlands Project: Conservation biology for the Continent. In *The Big Picture*, 14. Bellingham, WA.

Haas, Peter M. 1992. Introduction: Epistemic communities and international policy coordination. *International organization* 46, no. 1: 1–36.

Haines, Aubrey L. 1977. *The Yellowstone story: vol. 2*. Yellowstone National Park: Yellowstone Library and Museum Association.

Hansard. 1999a. *Official report of debates of the legislative assembly of Alberta: 3rd session, 24th legislature*. http://www.assembly.ab.ca/net/index.aspx?p=han&section=doc&v=3_24 [on lower menu, choose "Monday, December 6, 1999 1:30 PM"].

Hansard. 1999b. *Official report of debates of the legislative assembly of British Columbia, 1998/99 Legislative Session: 3rd session, 36th parliament*. May 6, Vol. 14, No. 25. http://www.legis.gov.bc.ca/hansard/36th3rd/h0506p9.htm.

Harvey, Ann. 1998. *A sense of place: Issues, attitudes and resources in the Yellowstone to Yukon ecoregion*. Canmore, Alberta: Yellowstone to Yukon Conservation Initiative. http://www.y2y.net/science/conservation/y2yatlas.pdf.

Hayes, Derek. 2001. *First crossing: Alexander Mackenzie, his expedition across North America, and the opening of the continent*. Seattle, WA: Sasquatch Books.

Hendee, John C., and Chad P. Dawson. 2002. *Wilderness management*. Golden, CO: Fulcrum Publishing.

Herrero, Stephen. 1998. Science and conservation in the Yellowstone to Yukon. In *A sense of place: Issues, attitudes and resources in the Yellowstone to Yukon ecoregion*, ed. Ann Harvey, 5–6. Canmore, Alberta: Yellowstone to Yukon Conservation Initiative.

Heuer, Karsten. 2000. A grizzly a day. *Backpacker* (April): 66–72+.

Heuer, Karsten. 2002. *Walking the big wild: From Yellowstone to the Yukon on the grizzly bears' trail*. Toronto: McClelland & Stewart.

Heuer, Karsten. 2004. *Walking the big wild: From Yellowstone to the Yukon on the grizzly bears' trail*. Seattle: The Mountaineers Books.

Hiss, Tony, and Christopher Meier. 2004. *H2O—Highlands to Ocean: A first close look at the outstanding landscapes and waterscapes of the New York/New Jersey Metropolitan Region*. Morristown, NJ: Geraldine R. Dodge Foundation. http://www.grdodge.org/environment_h2obook.htm.

Hummel, Monte, Sherry Pettigrew, and John Murray. 1991. *Wild hunters: Predators in peril*. Toronto, Canada: Key Porter Books.

Hutton, Carrie. 1998. Wildlife supporter to hike through Hinton. *Hinton Parklander*, October 14, A9.

Jessen, Sabine, and Natalie Lerch. 1999. Baja to Bering Sea: A North American marine conservation initiative. *Environments* 27, no. 3: 67–89.

Johns, David. 1998. A Y2Y history: Rationale & meeting summaries. In *Y2Y network handbook: Principles, policies and guidelines*, 10–13. Canmore, Alberta: Yellowstone to Yukon Conservation Initiative.

Johns, David. 2003a. Our real challenge: Managing ourselves instead of nature. In *Science and stewardship to protect and sustain wilderness values*. RMRS-P-27. Proceedings of the Seventh World Wilderness Congress Symposium, November 2–8, 2001, Port Elizabeth, South Africa, ed. Alan Watson and Janet Sproull, 172–176. Ogden, UT: U.S. Department of Agriculture, Forest Service, Rocky Mountain Research Station. http://www.fs.fed.us/rm/pubs/rmrs_p027/rmrs_p027_172_178.pdf.

Johns, David M. 2003b. The Wildlands Project outside North America. In *Science and stewardship to protect and sustain wilderness values*. RMRS-P-27. Proceedings of the Seventh World Wilderness Congress Symposium, November 2–8, 2001, Port Elizabeth, South Africa, ed. Alan Watson and Janet Sproull, 114–120. Ogden, UT: U.S. Department of Agriculture, Forest Service, Rocky Mountain Research Station. http://www.fs.fed.us/rm/pubs/rmrs_p027/rmrs_p027_114_120.pdf.

Jonkel, Chuck. 1988–89. Defining a bioregion: The Wild Rockies—A Glacier/Waterton view of the bioregion. *Columbiana* (Winter): 66.

Legault, Stephen. 1999. Heroes of this earth: Conversations with Wendy Francis and Harvey Locke. *Encompass* (April–May).

Levesque, Suzanne Marie. 2000. From Yellowstone to Yukon: Combining science and advocacy to shape public opinion and policy. Dissertation, University of California-Irvine.

Levesque, Suzanne Marie. 2001. The Yellowstone to Yukon Conservation Initiative: Reconstructing boundaries, biodiversity, and beliefs. In *Reflections on water: New approaches to transboundary conflicts and cooperation*, ed. Joachim Blatter and Helen Ingram, 123–162. Cambridge, MA: MIT Press.

Lewis, Meriwether, and William Clark. 1904 [2001]. *Original journals of the Lewis and Clark expedition, 1804–1806, Volumes I–VII*. New York: Dodd Mead & Company (reprinted by Digital Scanning, Inc., Scituate, MA).

Liidlii Kue First Nation, Canadian Parks and Wilderness Society–Yukon Chapter, and Canadian Parks and Wilderness Society–Northwest Territories Chapter. 1998. *A northern vision for the Yellowstone to Yukon Conservation Initiative*. Fort Simpson, Northwest Territories, Canada.

Locke, Harvey. 1994. Preserving the wild heart of North America. *Borealis* (Spring): 18–24.

Locke, Harvey. 1997. The role of Banff National Park as a protected area in the Yellowstone to Yukon mountain corridor of western North America. In *National parks and protected areas: Keystones to conservation and sustainable development*, ed. James Gordon Nelson and Rafal Serafin, ix, 292. New York: Springer, 117–124.

Locke, Harvey. 1998. Yellowstone to Yukon Conservation Initiative. In *Linking protected areas with working landscapes conserving biodiversity: Proceedings of the Third International Conference on Science and Management of Protected Areas, May 12–16, 1997*, ed. Neil Munro and J. H. Martin Willison, 255–259. Wolfville, Nova Scotia: Science and Management of Protected Areas Association.

Locke, Harvey. 2002. E-mail to Charles C. Chester. Files of Charles C. Chester, Cambridge, MA.

Locke, Sarah. 2000. *Yukon's Tombstone Range and Blackstone Uplands: A traveller's guide*. Whitehorse, Yukon: Yukon Chapter of the Canadian Parks and Wilderness Society.

Lowey, Mark. 1997. Yellowstone-to-Yukon trek highlights vast conservation plan. *Calgary Herald*, August 30, A8.

Mace, Richard D. 2004. Integrating science and road access management: Lessons from the Northern Continental Divide Ecosystem. *Ursus* 15, no. 1: 129–136.

Mackenzie, Alexander. 1801 [2001]. *The journals of Alexander Mackenzie*. Santa Barbara, CA: Narrative Press.

Mahr, Marcy, and Melissa Mauro. 2004. *Making science, making change: Celebrating five years of research and collaboration in the Yellowstone to Yukon region, 1999–2003*. Canmore, Alberta: Yellowstone to Yukon Conservation Initiative. Symposium Compendium, May 7–9, 2003, University of Calgary.

Marynowski, Susan. 1992. Paseo Pantera: The great American biotic interchange. Special issue, *Wild Earth*: 71–74.

Mayhood, Dave, Rob Ament, Rich Walker, and Bill Haskins. 1998. Selected fishes of Yellowstone to Yukon: Distribution and status. In *A sense of place: Issues, attitudes and resources in the Yellowstone to Yukon ecoregion*, ed. Ann Harvey, 77–91. Canmore, Alberta: Yellowstone to Yukon Conservation Initiative.

McCoy, Michael. 1998. *Journey to the Northern Rockies*. Old Saybrook, CT: Globe Pequot Press.

McMillon, Scott. 1998a. Hikers nurture idea of Rockies corridor: Yellowstone-to-Yukon wildlife road touted. *Denver Post*, June 28, B2.

McMillon, Scott. 1998b. Small group has big plans to save Rockies corridors: Couple will trek from Yellowstone to Yukon to publicize project. *Seattle Times*, July 5, A4.

Medeiros, Paul. 1992. A proposal for an Adirondack primeval. Special issue, *Wild Earth*: 32–42.

Merrill, Troy, and David Mattson. 1998. Land cover structure of Yellowstone to Yukon. In *A sense of place: Issues, attitudes and resources in the Yellowstone to Yukon ecoregion*, ed. Ann Harvey, 27–29. Canmore, Alberta: Yellowstone to Yukon Conservation Initiative.

Meyer, Judith L. 1996. *The spirit of Yellowstone: The cultural evolution of a national park*. Lanham, MD: Rowman & Littlefield.

Mitchell, Alanna. 1997. Building a home for the grizzlies to roam. *Globe and Mail*, October 6.

Mitchell, Alanna. 1998. Hikers aim to revive wildlife in Rockies. *Globe and Mail*, September 29.

Mitchell-Banks, Paul J. 2003. Protecting and sustaining wilderness values in the Muskwa-Kechika management area. In *Science and stewardship to protect and sustain wilderness values (Seventh World Wilderness Congress Symposium, November 2–8, 2001, Port Elizabeth, South Africa)*, ed. Alan Watson and Janet Sproull, Proceedings RMRS-P-27:172-176. Ogden, UT: U.S. Department of Agriculture, Forest Service, Rocky Mountain Research Station.

Muskwa-Kechika Information Office. 2005a. Management area. http://www.muskwa-kechika.com/management-area.

Muskwa-Kechika Information Office. 2005b. Protected areas. http://www.muskwa-kechika.com/management-area/protected.asp.

New England BioLabs, Inc. 2004. Room to room. In *2005–06 catalog & technical reference*. Beverly, MA. http://www.neb.com/nebecomm/biodiversity.asp.

Newman, Brownie, Hugh Irwin, Karen Lowe, Aimée Mostwill, Stephen Smith, and Jesse Jones. 1992. Southern Appalachian wildlands proposal. Special issue, *Wild Earth*: 46–60.

Newman, Peter C. 2003. We got there first! *MacLeans*, June 23. http://www.macleans.ca/switchboard/article.jsp?content=20030623_61158_61158#.

Newquist, Susan. 2000. The last great mountain ecosystem. *Backpacker* (April): centerfold.

Noss, Reed F. 1992. The Wildlands Project land conservation strategy. Special issue, *Wild Earth*: 10–25.

Pannell, Kerr & Foster. 1990. Trail of the great bear: Executive summary. Prepared for Alberta Tourism and Travel Montana.

Parks Canada. 1999. *National Parks mountain guide: The official Parks Canada visitor's guide*. Banff, Jasper, Kootenay, Yoho, Mount Revelstoke, and Glacier National Parks.

PEICNP. 2000. *"Unimpaired for future generations"? Protecting ecological integrity with Canada's national parks*. Ottawa: Panel on the Ecological Integrity of Canada's National Parks, Parks Canada Agency.

Peine, Mary Anne. 2000. The evolution of revolution: A brief history of environmental activism in the Northern Rockies. *Camas: The Nature of the West* (Spring-Summer): 14–17, 31–32.

Phillips, Mike. 2000. Conserving biodiversity on and beyond the Turner lands. *Wild Earth* 10, no. 1: 91–94.

Phillips, Mike. 2001. A private effort to conserve biological diversity. *Endangered Species Bulletin* (September): 26–29.

Posewitz, Jim. 1998. Yellowstone to the Yukon (Y2Y): Enhancing prospects for a conservation initiative. *International Journal of Wilderness* 4, no. 2: 25–27.

Quinn, Daniel. 1992. *Ishmael*. New York: Bantam/Turner Book.

Quinn, Daniel. 2005. *Ishmael* [website]. http://www.ishmael.com/Origins/Ishmael.

Rasker, Ray. 1997. Economic trends in the Yellowstone to Yukon region: A synopsis. In *Connections: Proceedings from the first Conference of the Yellowstone to Yukon Conservation Initiative, Waterton-Glacier International Peace Park, October 2–5*, ed. Stephen Legault and Kathleen Wiebe, 104–111. Canmore, Alberta: Yellowstone to Yukon Conservation Initiative.

Rasker, Ray. 1999. *People, commerce and the environment in the Yellowstone to Yukon region*. Whitehorse, Yukon: Canadian Parks and Wilderness Society, Yukon Chapter.

Rasker, Ray, and Ben Alexander. 1997. *The new challenge: People, commerce and the environment in the Yellowstone to Yukon region*. Bozeman, MT: Wilderness Society.

Rasker, Ray, and Ben Alexander. 2000. The changing economy of Yellowstone to Yukon: Good news for wild lands? *Wild Earth* 10, no. 1: 99–103.

Rasker, Ray, and Ben Alexander. n.d. *Economic trends in the Yellowstone to Yukon region: A synopsis, DRAFT (for the Yellowstone-to-Yukon Biodiversity Conservation Strategy Environmental & Cultural Resources Atlas)*. Files of the Yellowstone to Yukon Conservation Initiative, Canmore, Alberta.

Reeves, Brian O.K. 1998. Sacred geography: First Nations of the Yellowstone to Yukon. In *A sense of place: Issues, attitudes and resources in the Yellowstone to Yukon ecoregion*, ed. Ann Harvey, 31–50. Canmore, Alberta: Yellowstone to Yukon Conservation Initiative.

Reiger, John F. 1997. Wildlife, conservation, and the first forest reserve. In *American forests: Nature, culture, and politics*, ed. Char Miller, 35–47. Lawrence: University Press of Kansas.

Robbins, Jim. 1993. *Last refuge: The environmental showdown in Yellowstone and the American West*. New York: William Morrow.

Robinson, Bart. 1997. *The Yellowstone to Yukon Conservation Initiative: To restore and protect the wild heart of North America* [pamphlet]. Files of the Yellowstone to Yukon Conservation Initiative, Canmore, Alberta.

Rockwell, David B. 1998. *The nature of North America: A handbook to the continent: Rocks, plants, and animals*. New York: Berkley Books.

Ronda, James P. 1989. Dreams and discoveries: Exploring the American West, 1760–1815. *The William and Mary Quarterly* 46, no. 1: 145–162.

Sandwith, Trevor, Clare Shine, Lawrence Hamilton, and David Sheppard. 2001. *Transboundary protected areas for peace and co-operation*. Gland, Switzerland: World Conservation Union.

Sawchuk, Wayne. 1992. Letter to Rick Zammuto. Files of Rick Zammuto, Save-the-Cedar League, Crescent Spur, British Columbia.

Sawchuk, Wayne. 2004. *Muskwa-Kechika: The wild heart of Canada's Northern Rockies*. Chetwynd, British Columbia: Northern Images.

Schmitt, Catherine. 2005. On the shoulders of Mount A: Protecting southern Maine's coastal forest. *Northern Sky News*. October, 1+. http//www.northernskynews.com/backissue%20pages/Mount%20Agamenticus.html.

Schneider, Howard. 1997. Conservationists take stock of the land. *Washington Post*, October 27.

Schullery, Paul, and Lee H. Whittlesey. 2003. *Myth and history in the creation of Yellowstone National Park*. Lincoln: University of Nebraska Press.

Senge, Peter M. 1990. *The fifth discipline: The art and practice of the learning organization*. New York: Doubleday/Currency.

Service, Robert W. 1953. *The best of Robert Service*. New York: Perigree.

Shelford, Victor E. 1963. *The ecology of North America.* Chicago: University of Illinois Press.

Sherpa, Mingma Norbu, Sangay Wangchuk, and Eric Wikramanayake. 2004. Creating biological corridors for conservation and development: A case study from Bhutan. In *Managing mountain protected areas: Challenges and responses for the 21st Century,* ed. Harmon David and Worboys Graeme, 128–134. Italy: Andromeda Editrice.

Sherrington, Peter. 2005. Striking gold. *Birder's World* (August). http://www.birdersworld.com/brd/default.aspx?c=i&id=119.

Sifford, Belinda and Charles C. Chester. Forthcoming, 2006. Bridging conservation across La Frontera: An unfinished agenda for peace parks along the U.S.-Mexico divide. In *Peace parks: Conservation and conflict resolution,* ed. Saleem Ali. Cambridge, MA: MIT Press.

Snyder, Gary. 1992. Coming in to the watershed. Special issue, *Wild Earth*: 65–70.

Soulé, Michael. 1992. A vision for the meantime. Special issue, *Wild Earth*: 7–8.

Soulé, Michael, and Reed Noss. 1998. Rewilding and biodiversity: Complementary goals for continental conservation. *Wild Earth,* 18–28. http://www.twp.org/files/pdf/rewilding.pdf.

STCL. 1992. *Newsletter* (Fall). Crescent Spur, British Columbia: Save-the-Cedar League.

Stegner, Wallace. 1990. It all began with conservation. *Smithsonian* 21 (1): 34.

Stephens, Doug. 2004. Quabbin to Cardigan: The next landscape scale collaborative. *Views from Mount Grace* (Spring): 1, 3.

Struzik, Ed. 1998. A walk on the wild side. *Edmonton Journal,* October 17.

Struzik, Ed. 1999. Step by step. *Equinox* (February-March): 20.

Sunstein, Cass R. 2000. Situationism (book review of *The Tipping Point*). *New Republic* (March 13), 42.

Tabor, Gary M. 1996. *Yellowstone-to-Yukon: Canadian conservation efforts and continental landscape/biodiversity strategy.* Boston: Henry P. Kendall Foundation.

Tabor, Gary M., and Michael E. Soulé. 1999. *Yellowstone-to-Yukon.* Seattle, WA: Wilburforce Foundation.

TGB. 1998. Trail of the Great Bear. http://www.trailofthegreatbear.com/trail_index.html.

TGB. 2004–05a. *Trail of the great bear map.* Promotional map. Waterton Park, Alberta: Trail of the Great Bear.

TGB. 2004–05b. *Touring guide & Website directory.* Promotional pamphlet. Waterton Park, Alberta: Trail of the Great Bear.

Torgrimson, Christine. 2004. *A chronicle of successes.* Canmore, Alberta: Yellowstone to Yukon Conservation Initiative, April 26.

U.S. House. 1994. *The Northern Rockies Ecosystem Protection Act of 1993.* 103rd Cong., 2nd session. H.R. 2638.

U.S. House. 2005. *The Rockies Prosperity Act.* 109th Cong, 1st Session, H.R. 1204.

Waldt, Ralph. 2004. *Crown of the continent.* Helena, MT: Riverbend Publishing.

Warner, Gerry. 1998. A boy and his dog. *East Kootenay Weekly,* August 26, 4.

Webb, Trish. 1995. Yellowstone to Yukon Biodiversity Strategy Communications Plan. Files of the Yellowstone to Yukon Conservation Initiative, Canmore, Alberta.

Weisman, Alan, and Jay Dusard. 1986. *La frontera: The United States border with Mexico*. San Diego, CA: Harcourt Brace Jovanovich.

Wilburforce Foundation. n.d. Conservation Leadership Award. http://www.wilburforce.org/grantees/past_conservation.cfm.

Wildlands Project. 1999. *Questions, answers & information*. Tucson, AZ: Author.

Wilkinson, Todd. 2000. New pearl in emerald necklace. *Christian Science Monitor*, December 8. http://www.csmonitor.com/durable/2000/12/08/fp6s1-csm.shtml.

Willcox, Louisa. 1998. The wild heart of North America: A new perspective. In *A sense of place: Issues, attitudes and resources in the Yellowstone to Yukon ecoregion*, ed. Ann Harvey, 1–3. Canmore, Alberta: Yellowstone to Yukon Conservation Initiative.

Wolf, Alan. 2000. The next big thing. *New York Times*, March 5.

Wuerthner, George. 2001a. Keeping the grizzly in Grizzly Creek. *Wilderness*, 12–17. Annual publication of the Wilderness Society, Washington, DC.

Wuerthner, George. 2001b. A man for the bears. *Greater Yellowstone Report* (Early winter): 8.

Y2Y Biodiversity Strategy. 1996. Meeting minutes, Kananaskis, Alberta, April 13–14. Files of the Yellowstone to Yukon Biodiversity Strategy, Canmore, Alberta.

Y2Y Conservation Initiative. 1997a. Draft: *A sense of place: Issues, attitudes and resources in the Yellowstone to Yukon bioregion*. Summary of the Interim Draft, prepared for the Y2Y Connections Conference, October 2–5. Files of the Yellowstone to Yukon Conservation Initiative, Canmore, Alberta.

Y2Y Conservation Initiative. 1997b. Minutes of the Yellowstone to Yukon Conservation Initiative Council meeting, April 26–27, Kalispell, MT. Files of the Yellowstone to Yukon Conservation Initiative, Canmore, Alberta.

Y2Y Conservation Initiative. 1998. Y2Y Workplan: Overview. In *Y2Y network handbook: Principles, policies and guidelines*. Files of the Yellowstone to Yukon Conservation Initiative, Canmore, Alberta.

Y2Y Conservation Initiative. 2001. Our mission. Miistakis Institute for the Rockies. http://www.rockies.ca/Y2Y/overview/default.htm.

Y2Y Conservation Initiative. c.2001a. Yellowstone to Yukon [pamphlet]. Files of the Yellowstone to Yukon Conservation Initiative, Canmore, Alberta.

Y2Y Conservation Initiative. c.2001b. Y2Y Q&A 2001. Files of the Yellowstone to Yukon Conservation Initiative, Canmore, Alberta.

Y2Y Conservation Initiative. 2002. *Exploring common ground on the Dena Tuna: A Gathering of First Nations and Native Americans from the Yellowstone to Yukon Region, November 22–24, 2002, Y-Mountain Lodge, Banff, Alberta [Draft]*. Canmore, Alberta: Yellowstone to Yukon Conservation Initiative.

Y2Y Conservation Initiative. 2003a. *From the ground up*. Canmore, Alberta: Yellowstone to Yukon Conservation Initiative.

Y2Y Conservation Initiative. 2003b. Kaska, CPAWS sign historic agreement. *Connections* (Winter): 1–2. http://www.y2y.net/media/connections/winter03web.pdf.

Y2Y Conservation Initiative. 2003c. *Y2Y E-Update*. Canmore, Alberta: Yellowstone to Yukon Conservation Initiative, April.

Y2Y Conservation Initiative. 2004. The Y2Y vision. http://www.rockies.ca/Y2Y/overview/default.htm.

Y2Y Conservation Initiative. 2005. Y2Y Principles for working with aboriginal people. http://www.y2y.net/people/native.asp.

Y2Y Conservation Initiative. n.d.-a. Some facts about the Yellowstone to Yukon. In *Y2Y Interpreter's Kit*. Canmore, Alberta: Author.

Y2Y Conservation Initiative. n.d.-b. The Yellowstone to Yukon Conservation Initiative: Fiction vs. Fact [pamphlet]. Files of the Yellowstone to Yukon Conservation Initiative, Canmore, Alberta.

Y2Y Conservation Initiative. n.d.-c. *Y2Y Interpreter's Kit*. Files of the Yellowstone to Yukon Conservation Initiative, Canmore, Alberta.

Y2Y Coordinating Committee. 1996. Minutes of the Yellowstone to Yukon Coordinating Committee conference call, April 21. Files of the Yellowstone to Yukon Conservation Initiative, Canmore, Alberta.

Y2Y Coordinating Committee. 1999. Coordinating Committee meeting minutes, April 26–27, Kalispell, MT. Files of the Yellowstone to Yukon Conservation Initiative, Canmore, Alberta.

Y2Y Initiative. 1995. Working Group minutes. Files of the Yellowstone to Yukon Conservation Initiative, Canmore, Alberta.

Y2Y Network. 1997. Minutes of the Yellowstone to Yukon Network Coordinating Committee meeting. Files of the Yellowstone to Yukon Conservation Initiative, Canmore, Alberta.

Yukon Wildlands Project. 1996–97. *Wildlands vision*. Whitehorse, Yukon: Yukon Wildlands Project, Canadian Parks and Wilderness Society–Yukon Chapter, Yukon Conservation Society, Friends of the Yukon Rivers, and World Wildlife Fund–Canada.

Zakin, Susan. 1993. *Coyotes and town dogs: Earth First! and the environmental movement*. New York: Viking.

5

Conservation Effectiveness in the Territories of Chance

In Chapter 1, I described North America's borders as defining "territories of chance." No doubt, some historians would challenge the turn of expression. They might argue, for instance, that the geographical position of a certain river can at least partly dictate where a border is drawn. The point is well taken (if not wholly convincing to a fluvial geomorphologist, who might well equate surficial hydrology to dumb luck). But where diplomats agreed to extend straight-line borders through hundreds of miles of largely unexplored land, as in the cases of both the Sonoran Desert and the Northern Rockies, "territories of chance" applies.

The phrase also makes for an apt metaphor. Long after having chosen to look at the International Sonoran Desert Alliance and the Yellowstone to Yukon Conservation Initiative, I came upon my own territory of chance in the early-twentieth-century writings of the celebrated zoologist William T. Hornaday. Caustic and controversial, Hornaday was a conservation polymath whose wildlife campaigns have been credited with helping to save the American bison, pronghorn antelope, and northern fur seal from extinction (Dorsey 1998; Hornaday 1889; Shell 2002; Stroud 1985, 207). Early in his career as a conservationist, Hornaday published two companion books on the Canadian Rockies and the Sonoran Desert: *Camp-fires in the Canadian Rockies* (1906) and *Camp-fires on Desert and Lava* (c.1908). Both recount the daily activities of Hornaday and his companions in exploring the two territories, and the results are somewhat self-conscious travelogues with hints of boosterism and shades of anachronistic "Dear Reader" tediousness.

But these faults are just hints and shades. Overall, contemporary readers will find inspiration in Hornaday's eye for detail and his fixation on understanding wild animals such as the desert bighorn sheep and the grizzly bear. And today's conservationists will relish Hornaday's enlightened approach to hunting and trapping such big game. As he wrote in *Camp-fires in the Canadian Rockies*:

> For several reasons, I am totally opposed to the trapping of grizzlies for their skins, to poisoning them, and to permitting any hunter to kill more than one grizzly per year. In other words, I think the time has come to protect this animal, at least everywhere south of latitude 54°. . . . Eliminate the bears from the Canadian Rockies, and a considerable percentage of the romance and wild charm which now surrounds them like a halo, will be gone. (Hornaday 1906, 180)

And in an appendix to *Camp-fires in Desert and Lava* entitled "A Sportsman's Platform," the first of fifteen cardinal principles states that wildlife "is given to us *in trust*, for the benefit both of the present and future. We must render an accounting of this trust to those who come after us" (Hornaday c.1908, 355).

Less than a decade after writing those words, Hornaday came to the conclusion that if "wildlife is not saved through the initiative and the sacrifices of private individuals it will not be saved" (Hornaday 1914, 161). In this major treatise, *Wild Life Conservation in Theory and Practice*, he also at least twice reiterates the critical role of the individual in ensuring the conservation of wildlife:

> Private citizens and humanitarian organizations must not think that all the work and all the fighting for the saving of wild life should be done, or can be done, by the state game commissions. . . . In the protection of wild life, it seems to me that the average citizen does not even begin to realize his own power. I know it, and a great many other men know it, because we have seen the results that have been accomplished by the private citizen on the firing-line. (Hornaday 1914, 168, 170)

How did Hornaday come to the conclusion that—to put it in words that he might well eschew—civil society must save biodiversity? It is tempting to argue that his travels and experiences in the Northern Rockies and the Sonoran Desert were a major factor in his reasoning. Alas, reassuring as it would be to identify parallels, Hornaday's insight more likely came about because of the interminably contentious political battles he fought from his leadership positions at the New York Zoological Society and the Wildlife Committee of the Camp Fire Club of America—two of the country's earliest conservation NGOs. But imagine if you will, Dear Reader, this author's surprise in finding that a century ago, a founding father of the modern conservation movement had chosen to explore the same two regions I had—and that he came to the conclusion that it was private citizens who could be, perhaps must be, the agent of wildlife's continued presence on this continent. In the Ecclesiastical spirit of nothing new under the sun, this is the same conclusion I make here in relation to ISDA and Y2Y.

I do not, of course, pretend to the vitality or spirit of Hornaday. But whereas Hornaday made his assessments from the perspective of a political combatant, I have the luxury of looking at the role of civil society from a less invested and, dare I say it, more objective vantage point. And, again relying on contemporary language, assessing this role demands an examination of what it actually means to be effective in the world of biodiversity conservation.

The Multiple Meanings of Effectiveness

While both ISDA and Y2Y raise many questions about conservation and international cooperation, none is more important than that of effectiveness. Has either ISDA or Y2Y effectively enhanced conservation activities across an international border? Note that this question applies to both means and ends. For ISDA and Y2Y, the means are, in essence, various forms of cooperation; the ends are biodiversity conservation and sustainable development. People who give serious consideration to the issue of conservation effectiveness, and there are a growing number of them, rightly emphasize that the effectiveness of the means and the effectiveness of the ends are two quite distinct questions. This is readily apparent in the two case studies, in which effective cooperation did not always lead to effective conservation. Yet even as it is an important distinction to keep in mind, I have for the sake of simplicity and efficiency integrated the two in the following overview of conservation effectiveness.

In defining *effectiveness*, one must be wary of tautological entrapments. One might posit, for instance, that effectiveness is the "level of success or failure one has in implementing one's missions, goals, or objectives." That is perfectly true—but ultimately it does little more than semantically transfer the problem to figuring out what "success" and "failure" mean. And generically defining those terms may only keep us orbiting around the fundamental question. Furthermore, strange as it may initially sound, effectiveness is more than just a matter of achieving a stated goal. It is also a matter of knowing that one's actions are responsible for the achievement of a stated goal. As a highly simplified example, say you set out to protect a particular tract of forest from a timber company's plans to log it, and then for unrelated reasons the timber company goes out of business. It would be patently wrong in that case to say you have been effective in achieving your goal.

To get a sense of the inherent complexity in measuring the effectiveness of conservation activities, it helps to start with a bare-bones model of what ought to be entailed. In this simple formulation, effectiveness is defined as the degree to which conservation activities minimize or reverse the six generally accepted "proximate" causes of biodiversity loss: overexploitation (including persecution), habitat loss (e.g., agricultural expansion), habitat degradation (e.g., pollution and fragmentation), invasive species, climate change, and secondary effects. Note that the conservation literature offers numerous variations on this list: for three textbook alternatives, see Hunter Jr. (1996), Meffe and Carroll (1997), Wood et al. (2000, 5); for a popular rendition, see Wilson (2002); for an expanded version, see the "Taxonomy of Direct Threats" at the Web site of the Conservation Measures Partnership (http://www.conservationmeasures.org). Effectiveness can be modeled on the basis of these proximate causes:

$$E = f_i[O, L, D, I, C, S] + f_i[R]$$

Where:

E = Effectiveness of conservation activities
f_i = function of impact on
O = Overexploitation
L = Loss of habitat
D = Degradation of habitat
I = Invasive species
C = Climate change
S = Secondary effects
R = Ecological Restoration

Can this model shed light on the effectiveness of any particular conservation effort? Although it might serve as a helpful introduction to the myriad issues involved in biodiversity loss, the model's built-in oversimplifications limit its practical application in at least four ways. First, these threats significantly overlap. It can be difficult, for example, to distinguish the loss of habitat from the degradation of habitat, and it is not difficult to envision how climate change could cause or exacerbate most of the other threats. Second, as any conservationist is painfully aware, each individual factor in the model entails a staggering level of complexity, and to boil any one of them down to a mere letter for the sake of a model is arguably akin to misunderstanding the problem. The third limitation is that certain important factors are not made explicit in the model. One of those factors is the chronological change in *status* of whatever it is one is protecting—for instance, the population level of an endangered species both before and after conservation activities have occurred. Another missing factor concerns the place-based character of biodiversity conservation initiatives. Any place-based conservation activity will be heavily molded and shaped by the particular region's unique combination of characteristics. For instance, what is the biological significance of genetic variation and gene flow within the region? What attitudes do people in the region generally hold toward environmental causes and conservation NGOs? How dependent are local communities on extractive and land-consumptive industries—particularly in comparison to amenity-based industries? Although such characteristics could be captured in the above model, they are not readily apparent.

But the above limitations exemplify the generic types of complaints against almost any form of modeling exercise. The fourth and most substantive limitation of the model is that it focuses on *proximate* causes of biodiversity loss rather than on a myriad number of *root* causes. Ranging from population growth to ecological imperialism, the root causes of biodiversity loss have been defined as "the set of factors that truly drive biodiversity loss, but whose distance from the actual incidence of loss . . . makes them a challenge to identify and remedy" (Wood et al. 2000, 3). As with proximate causes, there are different ways to classify root causes. In their major treatise on the subject, Wood et al.(2000, 13) divided root causes of biodiversity loss into five "basic" categories: (1) demographic change; (2) poverty and inequality; (3) public policies, markets, and politics; (4) macroeconomic policies and structures; and (5) social change and development. This list is certainly imposing, and one can imagine why conservationists generally do not think about effectiveness at such an abstract level. But an even more abstract summation of root causes could be extracted out of the well-known I = PAT equation (Ehrlich and Ehrlich 1990; Ehrlich and Holdren 1971). Replacing the "I" for Impact with "E" for Effectiveness, the model would be:

$$E = f_i[P, A, T]$$

Where:

f_i = function of impact on
P = population
A = affluence
T = technology

Although thought provoking, practicing conservationists are likely to consider such models as little more than oversimplistic renderings of all too familiar phenomena (see Hynes 1993). But having gone this far in considering how to conduct a full accounting of effectiveness, we might as well finish our thought experiment with a model that takes into account some (not all) of the above limitations:

Stage 1: Assessment of the status of a conservation target (e.g., populations, species, habitat) and the degree to which that status deviates from a condition that is original, desired, appropriate, ideal, and so on.
Stage 2: Assessment of the extent of the proximate causes of biodiversity loss.
Stage 3: Assessment of the root causes of biodiversity loss.
Stage 4: Assessment of efforts to eliminate, minimize, or control the proximate and root causes of biodiversity loss.

Despite the elaboration, however, the level of abstraction here remains striking. Taking stage 1 alone, it is difficult enough to define either "ideal" or current states of certain "conservation targets." Generically, a growing number of researchers have been working to establish a set of status criteria under the rubric of "biodiversity indicators." Much of this effort has taken place under the aegis of the Convention on Biological Diversity (CBD) and the Global Biodiversity Forum, an ongoing forum hosted by NGOs, international organizations, and governments that usually takes place just before a meeting of the CBD or one of its subsidiary bodies. These indicators can be based either on process or on substantive outputs, and efforts have been made to establish criteria and indicators at a range of spatial scales, including at an ecoregional level (e.g., Gauthier and Wiken 1999; see Clark et al. 2002; World Conservation Monitoring Center 1992), at a national level (e.g., Ruggiero 1997; Turner et al. 1999), at a continental level (e.g., Pisanty-Baruch et al. 1999), and at the global level (e.g., Heywood 1995).

Yet even as investigators continue to work out biodiversity indicators at such broad scales, biologists have found it hard to flesh out indicators even at a relatively smaller scale. Take, for example, the long-standing vociferous debate over the status of the grizzly bear population in the Greater Yellowstone Ecosystem (GYE). Despite the significant amount of research conducted on the question, grizzly bear biologists have grappled for years with the lack of scientific data about something most people would probably assume is basic knowledge (see Mattson and Craighead 1994; Wilkinson 1998).

Getting a hold on such biological baselines is just stage 1; each subsequent stage only increases in complexity. For as difficult as it may be to determine the actual effect on the GYE's grizzly bear populations by a proximate threat in stage 2 (say, habitat fragmentation through road building in the GYE), it is an easier task than determining the effect of a root cause in stage 3 (say, human population growth in the GYE). Add to that already complex situation the severe practical problems that beset any attempt to measure organized human endeavors, including the lack of accountability, the desire to continue a program after its completion (whether successful or not), and the fear of "hard-nosed" evaluation policies undermining altruistic goals. Given all those costs and impediments, who dares try to measure effectiveness? The daunting prospect of doing so helps explain the common adage that the conservation community is inordinately preoccupied with "putting out the brushfires" of an unceasing litany of threats—and that consequently it dares not whittle away scarce resources by measuring the effectiveness of past conservation activities.

All this inherent complexity requires conservationists to accept the somewhat humbling premise that effectiveness is contextual; from "saving the whales" to "saving one acre of forest," each conservation initiative must to some degree develop its own criteria for effectiveness. Just as real estate agents are fond of noting *the* three factors involved in setting property values ("location, location, and location"), a similar list might be drawn up for measuring effectiveness ("context, context, and context"). One might even say that the diversity of "effectiveness measurements" should reflect the very diversity of genes, species, and ecosystems that conservationists are trying to protect.

I so heavily emphasize the importance of context because it is fundamental to understanding conservation effectiveness. In fairness, however, many of those on the forefront of thinking about conservation effectiveness would likely say that I have overemphasized the point and that it is hardly an impossible task to assess effectiveness. As will become evident, conservationists and their observers can be systematic when it comes to measuring effectiveness and have even built up a sizable body of analysis on conservation effectiveness. For although the adage about "putting out brushfires" still resonates, conservationists really do ponder and fret over whether all their work has made any real difference in changing the world. Increasingly, they decry the dearth of systematic analysis of their own effectiveness as myopic and ultimately self defeating. The less critically we assess our effectiveness, it is now almost commonplace to hear within the conservation community, the less likely we are to achieve success in current and future endeavors (Christensen 2003; Kleiman et al. 2000, 363; Margoluis and Salafsky 1998, 7). And despite the innumerable difficulties involved, conservationists have devised a number of practical methods for assessing conservation effectiveness.

Effecting Effectiveness

Conservationists have devised at least three broad approaches to measuring effectiveness: (1) project and program-oriented evaluations, (2) organizational assessments, and (3) international regime models. Although the first two are limited in their scope and the utility of the third remains untested, it is worth a short look at the feasibility of applying each of these lenses to ISDA and Y2Y. What will become apparent is that, as described in the previous two chapters, there is no easy answer to the question of what ISDA and Y2Y *are*, and that consequently it is difficult to pin down which approach to use in assessing their effectiveness.

Project and Program-Oriented Evaluations

The conservation literature contains numerous project and program-oriented approaches to the issue of effectiveness. *Projects* have been defined as "any set of actions undertaken by any group of managers, researchers, or local stakeholders interested in achieving certain defined goals and objectives" (Margoluis and Salafsky 1998, 7), while *program* is generally interpreted as a collection of thematically linked projects over a relatively longer time frame. *Evaluation* and *monitoring* constitute two key terms in the relevant literature, much of which offers cradle-to-grave project life-cycle analyses—or "how-to" guidelines for program and project managers. One of the more promising concepts of this growing body of literature has been the "threat reduction assessment" approach, which monitors changes in the level of threats as "a proxy measurement of conservation success" (Salafsky and Margoluis 1999, 833).

A large portion of this literature emanated out of the Biodiversity Support Program (BSP), a former consortium of the World Wildlife Fund, The Nature Conservancy, and the World Resources Institute that was funded by the U.S. Agency for International Development. (Although the BSP is now over, its sizable published output has been archived at http://www.worldwildlife.org/bsp/publications/index.html.) Several of those involved in the BSP went on to found Foundations of Success (FoS), a conservation organization that is spearheading the multi-NGO Conservation Measures Partnership and provides a vast storehouse of information on improving conservation practices (http://fosonline.org).

As Johns (1999) has commented, this general project-oriented approach to effectiveness "does not really address typical top-down, government-directed policy implementation." Nor does it directly address typical grassroots, NGO-directed networking—as in the cases of ISDA and Y2Y. That is, although both ISDA and Y2Y have implemented and participated in specific projects and programs, neither "project" nor "program" comes close to an adequate description of either initiative in its entirety. Kleiman et al. (2000, 358–359) might as well have been referring directly to ISDA or Y2Y when, after listing myriad recommendations for assessing programmatic effectiveness, they state, "Conservation programs involving several geographic or political areas (different states, different nations, different continents) may be especially difficult to review because the more agencies and actors involved, the more complex the organization and process."

Organizational Assessments

Organizational assessments consist of a number of approaches to improving the internal administration of a nonprofit organization. One significant component of this literature consists of examining the flow of financial resources, particularly the percent of an organization's budget being used for programmatic work as opposed to administration and salaries. At least one business publication, *SmartMoney Magazine* (Kadet 2000), has devised an efficiency scale for nonprofits based on such financial data, and several Web-based organizations have been established to assess charitable organizations (see, for example, www.charitynavigator and www.give.org).

Yet such analyses can be faulted both for relying on potentially distorted data and for not taking into account the numerous special circumstances facing nonprofits (e.g., capital campaigns, institutional capacity strengthening measures, internal strategic planning phases, and other organizational activities that may last multiple years). As the more reserved *Harvard Business Review* proclaimed early on—perhaps overstating the case—a "nonprofit organization's financial reports reveal absolutely nothing about its progress" (Harvey and Snyder 1987, 4). Rather, the *Review* advises: "Smart nonprofits are finding that goal-oriented management, combined with yardsticks to measure progress, may mean the difference between success and failure" (Harvey and Snyder 1987, 2). For the distinct organizational entities at the core of both ISDA and Y2Y—that is, their centralized staff and offices—such advice is no doubt worthwhile. Nonetheless, even the most thorough assessment of their central offices would hardly scratch the surface of the multifarious activities that have taken place under ISDA and Y2Y, and consequently this approach is limited as well.

International Regime Models

Over the past decade, a community of researchers has been making a systematic examination of the effectiveness of international environmental regimes (IERs). The now classic definition of an *international regime* describes it as a set of "implicit or explicit principles, norms, rules, and decision-making procedures around which actors' expectations converge in a given area of international relations" (Krasner 1983, 2). Perhaps the most widely recognized IER is the climate change regime, which in less formal terms can be described as all of the scientists, advocates, diplomats, agency officials, private-sector interests, and others who in some way are working to affect the way the international community responds to the problem of climate change. IER scholar Oran Young (and his colleagues) have deconstructed the broad idea of IER effectiveness into six "behavioral models"—perhaps better described as indicators of effectiveness—that "can help us identify and explain the pathways through which regimes affect collective outcomes" (Young and Levy 1999, 21–25). Framed as questions, the six indicators are as follows:

- Is the regime a *utility modifier*? Does it change the "utilitarian calculus" of individual actors?
- Does the regime *enhance cooperation*? Does it alleviate impediments to collective decision making?
- Is the regime a *bestower of authority*? Does it legitimize the existence of authoritative rules?
- Does the regime act as a *learning facilitator*? Does it increase the amount of factual information, thus changing "prevailing discourses, and even values"?
- Is the regime a *role definer*? Does it cause actors to "take on new roles" even when they have entered into the regime with "basic identities" and interests?
- Is the regime an *agent of internal realignment*? Has it changed the relative position of specific entities (such as corporations and environmental NGOs) within collective entities (such as states)?

Young (1999) and his colleagues fruitfully apply these six indicators to three IERs (international vessel-source oil pollution, the Barents Sea fisheries, and acid rain in Europe and North America). Could they be similarly applied to international initiatives such as ISDA and Y2Y? Although scholars of IERs have widely recognized the prominent role of civil-society actors in international environmental cooperation, they have nonetheless focused their analyses on IERs as primarily inter*governmental* processes. However, I have argued elsewhere that ISDA and Y2Y constitute a new form of IER (Chester 1999, 2003), and that there is tremendous potential for applying the lessons of this field to these two and similar initiatives. For example, a cursory overlay of these indicators on top of ISDA and Y2Y reveals that at least two of them are obviously applicable. First, both ISDA and Y2Y significantly enhanced cooperation between nongovernmental actors (and to some degree in ISDA's case, governmental actors, as well). Second, both have acted as learning facilitators, particularly between different cultural groups—whether that difference be as disparate as that between Tohono O'odham and "Anglos," or as relatively subtle as that between the Canadian and U.S. conservationists. The other four indicators might be applicable as well—for instance, one could argue that ISDA and Y2Y changed the "utilitarian calculus" of some NGOs in deciding where to focus their geographic or programmatic energies. In sum, further direct application of Young's six "behavioral models" would likely produce useful insights into the effectiveness of ISDA and Y2Y. Hopefully, as investigations into the effectiveness of IERs continue to grow, nongovernmental initiatives such as ISDA and Y2Y will be included in that broad research agenda.

A Comparative Approach to Effectiveness

In pondering how to assess the effectiveness of ISDA and Y2Y, I set aside the above approaches and turned instead to the two divergent methodological traditions of the fields of history and political science (though note that in practice, historians and political scientists utilize innumerable methodological approaches that would defy being swept into either of the two simplified camps I sketch here). On the one hand, I held to the historian's conviction that true knowledge results from in-depth inquiry into how events unfolded at a particular place and time. To get at the question of ISDA's and Y2Y's effectiveness, historians will assert, requires an individualized accounting of what each has actually accomplished in terms of particular circumstances and goals—again, "context, context, and context." This book's emphasis on the two case studies reveals my bias for the historical tradition. Were ISDA and Y2Y effective? Chapters 3 and 4 provided complex answers to this question by recounting what each initiative accomplished and did not accomplish.

Political scientists are less enthusiastic about such a particularistic approach, often disparaging it as mere fact accumulation rather than systematic thinking. Although no political scientist would wholly discount the importance of context, they generally assert that it is only through a comparative approach that we can begin to acquire a comprehensive understanding of why and how things happen. Thus, political scientists tend to prefer analyses that include a broad swath of case studies—or a "large N," that particular letter signifying the academically universal shorthand for "number of cases." Accordingly, coming from a political scientist, an "N of 1" is often heard a disparaging remark. Happily enough, I happen to have an N of 2, and it seems perfectly reasonable to conduct a comparative analysis of the different levels of effectiveness between ISDA and Y2Y.

But it is difficult to stifle the historian. ISDA and Y2Y have been radically different in circumstances, conceptualization, operation, and implementation—so different that any attempt to place them on a simplistic ordinal scale of effectiveness could preclude any real understanding of what they have or have not accomplished. Therefore, extreme caution must be taken not to fall into the trap of comparing the results of ISDA and Y2Y without taking into consideration their initial and ongoing conditions. The trap might look like this: "Here are two transborder initiatives that were both started by nongovernmental actors during the early 1990s, but whereas Y2Y took on a life of its own and is now found far beyond the confines of conservationists working under its aegis, ISDA remains a singularly identifiable entity based in the town of Ajo, Arizona." No doubt, more people today are aware of Y2Y than of ISDA, Y2Y has attracted more conservation funding than ISDA, and more networking has occurred under Y2Y than under ISDA. But, again, this is a patently unfair comparison because it does not take into account the drastically different circumstances that each initiative faced over the course of a decade.

How, then, does one avoid this trap and still garner the benefits of a comparative analysis? One way is to turn to the analytical tools of independent and dependent variables. For those fortunate enough to have avoided these shrouds of academic jargon, the terms warrant a quick explanation. A *dependent variable*, on the one hand, *requires explanation*; it is what a researcher is trying to figure out. *Independent variables*, on the other hand, *explain*; they are what a researcher hopes to identify as causing a particular value of a dependent variable. It is also important to understand that both dependent and independent variables can take on two or more values—a *value* in this context meaning a certain condition (or state of being) rather than a judgment of worth.

While this may seem straightforward, in the real world it can be nearly impossible to sort out what is a truly dependent variable from the vast universe of potential independent variables. This is particularly so when a dependent variable is characterized by so loose and open a concept as "effectiveness." In this regard, Young (1996, 8) points out that the "more we bear down on the specification of effectiveness as a dependent variable, the clearer it becomes that we are really dealing with a suite of related variables or, at best, a multidimensional variable whose separate dimensions need not and frequently do not co-vary in any simple way."

Keeping such warnings in mind, using dependent and independent variables allows for taking into account the initial and ongoing conditions of ISDA and Y2Y while retaining the benefits of a comparative analysis. And in the time that I have spent looking at the two initiatives, at least seven particular independent variables appear to have had significant causative influence on the dependent variable of effectiveness: mission breadth, constituency inclusion, communication systems, scientific participation and support, leadership dynamics, political backlash, and landscape vision.

Mission breadth can be thought of as the number of primary goals (or objectives, depending on whose management lingo one appropriates) that are subsumed under the broad mission of each initiative. Both ISDA and Y2Y included biodiversity conservation as well as human well-being in their mission statements. Yet from the start, ISDA's mission focused on enhancing multicultural communication for a number of primary goals, only one of which was to protect the region's biodiversity. In contrast, Y2Y's original mission was nearly entirely focused on biodiversity conservation. Substantive attention to sustainable communities and "human dimensions" were added somewhat later, and while those aspects have since become integral to the Y2Y vision, biodiversity conservation remains the overarching premise of Y2Y. Thus, although Y2Y was focusing on a much larger landscape than ISDA, the breadth of its mission was far less encompassing and it was thus arguably far less complex than that of ISDA. This is not to assert that biodiversity protection constitutes a simple problem—any conservationist working to avert species extinction and ecosystem degradation would justifiably chafe at the thought. But it is to say that Y2Y started out by picking up a piece of the larger "sustainable development" puzzle, whereas ISDA began by picking up the puzzle box and asking: Who wants to help put this thing together?

The answer to that question pertains to the next independent variable of *constituency inclusion*, which can be loosely defined as the diversity of "stakeholder interests" directly participating within each initiative. Early direct participation in the Y2Y Biodiversity Strategy, as it was first entitled, was limited to conservationists, scientists, and a smattering of others (e.g., foundation representatives). While the Y2Y Conservation Initiative has since labored to bring in representatives from a number of other stakeholder groups—such as sportsmen, ranchers, and business people—Y2Y remains at its core and in its various manifestations a conservationist-driven initiative. ISDA, on the other hand, began on the premise that all stakeholder groups from the region were to be included. Most important, the single most distinguishing difference in constituency inclusion was indigenous participation. Long before ISDA came about, both Hia-Ced and Tohono O'odham representatives had been working to assert their fundamental role in any organization that was going to claim to represent the region. Y2Y, in contrast, never claimed to be representing First Nation or Native American interests. Although extensive collaboration with indigenous groups in the Yukon occurred early on under the Y2Y aegis, it was not until 2003 that a tribal representative was brought onto the Coordinating Committee/Board, that a set of "Y2Y principles for Working with Aboriginal People" was established, and that a "historic" agreement was reached with three traditional peoples in the northern part of the Y2Y region (Y2Y Conservation Initiative 2003, 2005).

Directly related to constituency inclusion was the independent variable of *communication systems*, which were dramatically different between ISDA and Y2Y. Most obvious and significant, ISDA was daily confronted with the language barriers between O'odham, Spanish, and English. These barriers pervaded everything from person-to-person networking to staff efficiency. But the barriers were not only linguistic. Whereas most Y2Y participants had started using electronic communications before 1997 and the Y2Y electronic mailing list (EML) had been the foundation of Y2Y's networking capacity, the use of electronic communications such as email, EMLs, and Web sites within ISDA has been comparatively minimal (for instance, ISDA only established a Web site at the end of 2001 and has never used an EML). ISDA participants often found it difficult to communicate even by fax or mail due to poor communications and postal infrastructure. These problems hampered the networking potential of ISDA and swallowed up a great deal of staff and board time and resources.

Both ISDA and Y2Y began with *scientific participation and support*, and both subsequently found that most scientists were unwilling (sometimes unable) to spend a great deal of time sitting through meetings. But the similarity stops there, as scientific research continued to play a much more extensive role in the development of Y2Y than it did in the development of ISDA. For even as direct participation in Y2Y activities by scientists declined, other participants remained actively dedicated to supporting science for both internal and external decision making. Indeed, by maintaining a focus on science, Y2Y kept in its sphere several well-respected and influential scientists whose backing lent credence both to the idea and to the institutional structure of Y2Y. In contrast, the group of scientists who were at the originating 1992 Land Use Forum rarely attended the follow-up meetings that led to the foundation of ISDA. For the most part, the scientists who remained active in conservation issues in the

Sonoran Desert subsequently chose to attend meetings of the more activist Sonoran Desert Taskforce.

Strong *leadership dynamics* were instrumental in the development of both ISDA and Y2Y. However, while the leaders in both cases were highly articulate, energetic, and dedicated, the leadership of Y2Y remained both more consistent and more persistent than that of ISDA over the course of the 1990s. Most important, the changeover in the staff position of coordinator at ISDA led to an organizational vacuum that sucked the wind out of ISDA's sails at a critical time. While Y2Y saw its share of infighting and staff changeover due to leadership conflicts—and even saw a branch of the conservation movement in the U.S. Northern Rockies make a conscious break from Y2Y—it never experienced such an emotionally wrenching and institutionally debilitating conflict.

While both ISDA and Y2Y received considerable *political backlash*, it came in two different forms. For Y2Y, the backlash was vociferous and was aimed directly at anything with a Y2Y label. For ISDA, the backlash was aimed not so much at the organization as at what the organization was trying to do—namely, establish a biosphere reserve, and an *international* biosphere reserve at that. This difference had a somewhat ironic effect. While the political backlash against the Y2Y label did become problematic for individual conservationists and NGOs working within specific subregions in the Northern Rockies, it has arguably promoted the visibility and cachet of Y2Y on the regional, national, and international scales—which ultimately served to enhance Y2Y's public stature. In the Sonoran Desert, however, the broad political backlash against the goal precipitated the demise of what had become ISDA's signature applied endeavor.

Early in their respective histories each initiative defined its *landscape vision* geographically on a map, and from 1993 to today neither map has been significantly modified. On the one hand, ISDA's map circumscribed a region that was originally characterized as the "western Sonoran Desert," a term that has since fallen out of use with no currently accepted equivalent. In contrast, the label of Y2Y's map was "the Yellowstone to Yukon region," a geographic label that has stuck—at least as measured by the increased reference to the "Y2Y region" over the initiative's history. So whereas Y2Y offered an identifiable label for its vision of the landscape, nothing similar came out of ISDA. Although ISDA participants—particularly the O'odham—certainly had a clear vision of the landscape, that vision never took on a life of its own outside of ISDA or the region's indigenous communities, as was the case for Y2Y.

Arguably, this comparison is unfair because it is based on the circumstances (one might say dumb luck) of place and national self-identification. Conservationists in the Northern Rockies carved out a biogeographically coherent region that was bounded by two national icons, whereas such iconography lay in neither label of "western Sonoran Desert" nor "international Sonoran Desert." But this hardly means that the Sonoran Desert was devoid of its own meaningful symbols, and one cannot help but wonder whether ISDA could have found an identifiable landscape icon—even one not as widely resonant as "Yukon" or "Yellowstone." One must tread carefully here, for any recapitulation of the Y2Y model in the Sonoran Desert—say, G2G as a haphazard shot in the dark for "Goldwater to the Gulf"—would stretch even the most wonkish of imaginations (few people, after all, are aware that one of the country's largest bombing ranges was named for the former senator). However, what does immediately come to my mind is the potent icon of the saguaro cactus, which, despite being one of the most easily identified plants in the world, is unique to the Sonoran Desert. And beyond the saguaro, is it too far-reaching to suggest that the terrain and character of the Sonoran Desert have other features that could be molded into a landscape vision? As but one neighboring model, look to the east of the Sonoran Desert, where conservationists have helped define a bilateral region through the metaphor of "sky islands," a term now increasingly recognized both inside and outside of conservation circles (Foreman et al. 2000). Of course, hindsight makes it easy to suggest that ISDA could have created a more engaging vision. But regardless of what it might have done in this regard, there remains a deep contrast between the landscape visions of ISDA and Y2Y.

In sum, each of these seven independent variables not only helps explain the difference in effectiveness between the two initiatives, but also identifies the circumstances and conditions that caused that difference. The list is certainly incomplete, and it is important to keep in mind that these independent variables were identified through a comparative analysis of what "insiders" told me about their respective initiatives. It is, I hope, a valuable approach, but it is hardly the only one.

One final note from a comparative perspective: In talking with numerous ISDA and Y2Y participants, my questions regarding effectiveness elicited two fundamentally contrasting types of responses. When ISDA participants spoke about effectiveness, they mostly focused on institutions and chronologies—which is to say that they gave me a *historical narrative* of the programmatic effort to create ISDA as a sustainable enterprise, a major component of which was the goal of creating an international biosphere reserve. Most Y2Y interviewees, in contrast, offered a *description of an idea*, defining effectiveness in terms of the power of the "Y2Y vision." They argued that Y2Y had changed not only the way they thought about the region, but also how they actually went about their work—thus improving the landscape of international cooperation for biodiversity protection in the Northern Rockies.

Enough of differences. Harkening back to Hornaday's observations, both ISDA and Y2Y were started by people who were concerned about what was happening to the larger landscape (particularly those expansive tracts of public lands in both regions) but who did not work for the government. Composed of individual citizens reaching across international borders, both initiatives consciously ignored the traditional government monopoly over international cooperation. They shared, in other words, the innovative trait of working not just to influence government action—something long familiar on the international scene—but also to understand and cooperate with each other beyond the confines of governmental approval and protocol. As such, ISDA and Y2Y stand as useful models—or at least, practical lessons—for future approaches to transborder conservation in North America and beyond.

Civil Society versus Governments?

Long before I heard of ISDA or Y2Y, I was interested in how the international community could most effectively protect biodiversity. By "international community," I specifically had in mind government officials, as well as the scientists and advocates who incessantly (and sometimes successfully) told those government officials what to do. Thus blindered by such a limited interpretation of *international*, what immediately impressed me about ISDA and Y2Y was their focus on large landscapes that crossed international borders—and not the fact that both arose out of civil society. But after I did only a bit of rummaging around, it became manifestly clear that ISDA and Y2Y were as much about cooperation between civil society actors as they were about large landscapes.

So what exactly is *civil society*? Although an academic cottage industry devoted to the subject has generated multiple nuanced answers to that question, the phrase is generally defined as "that arena of social engagement existing above the individual yet below the state" (Wapner 1996, 4). An even more common interpretation equates civil society with NGOs, often working together for some common purpose. As archetypal examples of "international civil society," ISDA and Y2Y made me wonder whether transborder cooperation among civil society actors might have as great an impact on conservation—or possibly even greater—than could cooperation among governments.

I was not asking the question in a historical vacuum, for a wing of the aforementioned cottage industry has long focused on the role of civil society in international affairs. To summarize briefly—and to simplify tremendously—scholars in this field have gone through three broad phases in their thinking. First, traditional theories of international relations held that only the governments of independent nation-states could be regarded as significant actors in the international arena. Civil society, from that "realist" perspective, has played a largely tangential role in international affairs. In the second phase, a group of scholars challenged that basic tenet, arguing that civil society had become a veritable source of power in international relations. Those scholars argued that whereas conservation-oriented NGOs had been working to influence international affairs since at least the start of the twentieth century (Hornaday's New York Zoological Society and Camp Fire Club, for instance), the past two decades have seen them rise to considerable influence in international affairs. These NGOs have garnered such power by lobbying governments to enter into environmental treaties, by monitoring the enforcement of treaties, by safeguarding biologically valuable habitats from destructive forms of development, and by educating the public on the need for international conservation. From this vantage point, NGOs appeared to be a significant force in driving the international conservation agenda—even to the degree of supplanting governments in certain roles.

In the third phase, mainstream scholars recognized much of the substance of these claims, accepting that through their large memberships, articulate voices, and political acumen, conservation NGOs have attained a well-respected role—or at least a well-dreaded role—in both domestic and international debates over conservation policy. But even as civil society actors have become highly influential in the international arena, from this third perspective they have yet to mount any real threat to the fundamental solidity of the state-centric system of international relations. In short, scholarship had arrived at the commonsense notion that the influence of civil society and the power of government were hardly mutually exclusive—and that the two reinforce each other in the constant interplay of conflict and cooperation (for a helpful description of this evolution, see Lipschutz and Mayer 1996, 8).

In the cases of ISDA and Y2Y, this type of dynamic relationship between civil society and government was not always apparent. Both initiatives were children of civil society, and both seemed to operate at times as if governments hardly existed. Nonetheless, governmental prerogatives permeated nearly every aspect of their goals. This was more obvious in ISDA, which early on established a short-lived governmental advisory board, received its first grant from the U.S. government, and eventually took on the challenge of convincing both federal governments to approve the idea of an international biosphere reserve. The situation was different for Y2Y, where government officials have only occasionally stepped into its meetings. Nevertheless, all Y2Y participants were fully cognizant that many endangered species in the lower forty-eight remain directly dependent on governmental protection, and that maintaining healthy wildlife populations in both countries would prove impossible without government support.

In short, ISDA and Y2Y were engaged in a regional version of what political theoretician Paul Wapner has called "world civic politics," in which environmental NGOs go beyond traditional lobbying efforts to work "underneath, above, and around the state to bring about widespread change" (Wapner 1996, 9). Wapner describes how three conservation NGOs—specifically Greenpeace, World Wildlife Fund, and Friends of the Earth—have achieved substantial conservation successes *without* working directly with or against governments. His essential argument, however, is not that governments were unimportant in regard to the ultimate goals of those three NGOs, but rather that each of those NGOs found its own way of influencing governments beyond the traditional approach of direct advocacy.

Both ISDA and Y2Y have conducted the bulk of their work with an approach similar to Wapner's three NGOs, but with the added complexity of having a comparatively lower level of internal organizational structure —a complexity that resulted both from ISDA's and Y2Y's network origins and from the constant need for internal "diplomacy" among their participants. But no matter how structurally complex ISDA and Y2Y were internally, neither could ignore the bonds connecting them to the effects of government decision making. Consequently, understanding

ISDA and Y2Y does not demand an answer as to whether and how they worked to influence government decision making. Rather it requires asking: What were the advantages of working outside the realm of traditional diplomatic venues?

For ISDA, the most distinct advantage of having roots in civil society was that such a status enabled it to be the one external agent able to bring together both citizens and public land managers from throughout the region. Indeed, long before ISDA was formed, many public land managers understood that accomplishing their individual missions would require cooperation across international and agency borders—but they did not have the requisite external impetus to tear down the solid bureaucratic walls keeping them apart. ISDA provided that impetus, and although the ultimate goal of an international biosphere reserve was never achieved, a far higher level of international and interagency communications was the actual result.

For Y2Y, the advantage can be summarized in one word: empowerment. By changing the way conservationists fundamentally conceived of the landscape and bringing new resources into the region, Y2Y empowered conservationists to work with or fight against government agencies from a much more coordinated and strengthened position revolving around a common vision. Although it remains difficult to identify the direct causal links from Y2Y to governmental decision making, there can be little doubt that conservationists garnered much greater influence through the complex mix of inspiration, networking, and funding brought about by Y2Y.

For ISDA and Y2Y participants who have read *Democracy in America* by Alexis de Tocqueville, all of this will sound a familiar chord. For among his many insights into the American character, Tocqueville may be best remembered for his thoughts on civil society in the United States:

> As soon as several of the inhabitants of the United States have conceived a sentiment or an idea that they want to produce in the world, they seek each other out; and when they have found each other, they unite. From then on, they are no longer isolated men, but a power one sees from afar, whose actions serve as an example; a power that speaks, and to which one listens. (Tocqueville 2000, 492)

I hesitate to recite that quotation; Tocqueville was certainly not thinking of either Mexico or Canada when he published those words in 1840 ("The Union has no neighbors," he wrote [Tocqueville 2000, 265]). And were he reincarnated to revisit the whole of North America today, he still might not feel compelled to describe either in equally emphatic terms. But what he might say is that along with the cultural exports of Hollywood, democratic institutions, Coca-Cola, and various forms of high technology, the United States has managed to export to its two geographic neighbors the idea—the ideal—of civil society. This is evident with the rise of civil society in Mexico in response to the global movement toward economic integration (Wise et al. 2003). And the export of civil society has been even more welcome in the United States' northern neighbor—and thus we find Y2Y as an export *from* Canada *to* the United States. Overall, then, Toqueville's thoughts have become tentatively continental, for across both of the major borders dividing North America, conservationists have sought each other out, they have united, and they have engendered enduring examples of the "power that speaks, and to which one listens."

References

Chester, Charles C. 1999. Civil society, international regimes, and the protection of transboundary ecosystems: Defining the International Sonoran Desert Alliance and the Yellowstone to Yukon Conservation Initiative. *Journal of International Wildlife Law and Policy* 2, no. 2: 159–203.

Chester, Charles C. 2003. Biodiversity over the edge: Civil society and the protection of transborder regions in North America. Ph.D. diss., Fletcher School of Law and Diplomacy.

Christensen, Jon. 2003. Auditing conservation in an age of accountability. *Conservation in Practice* (Summer): 12–19.

Clarke, Robin, Robert Lamb, and Dilys Roe Ward. 2002. *Global environmental outlook 3*. London: Earthscan Publications Ltd. for the United Nations Environment Programme.

Dorsey, Kurkpatrick. 1998. *The dawn of conservation diplomacy: U.S.-Canadian wildlife protection treaties in the progressive era*. Seattle: University of Washington Press.

Ehrlich, Paul R., and Anne H. Ehrlich. 1990. *The population explosion*. New York: Simon & Schuster.

Ehrlich, Paul R., and John P. Holdren. 1971. Impact of population growth. *Science* 171, no. 3977: 1212–1217.

Foreman, Dave, Mike Seidman, Bob Howard, Jack Humphrey, Barbara Dugelby, and Andy Holdsworth. 2000. The Sky Islands Wildlands Network: Diverse, beautiful, wild, and globally important. *Wild Earth* 10, no. 1: 11.

Gauthier, David A., and Ed B. Wiken. 1999. Reporting on macro-ecosystems: The Great Plains of North America. *George Wright Forum* 16, no. 2: 52–63.

Harvey, Philip D., and James D. Snyder. 1987. Charities need a bottom line too. *Harvard Business Review* 66: 1 (January/February), 2–6.

Heywood, Vernon H. 1995. *Global Biodiversity Assessment.* New York: Cambridge University Press for the United Nations Environment Programme.

Hornaday, William Temple. 1906. *Camp-fires in the Canadian Rockies.* New York: C. Scribner's Sons.

Hornaday, William Temple. c.1908. *Camp-fires on desert and lava.* London: T. Werner Laurie (originally published 1908, C. Scribner's Sons; 1983 reprint by University of Arizona Press).

Hornaday, William Temple. 1914 [1972]. *Wild life conservation in theory and practice: Lectures delivered before the Forest School of Yale University, 1914.* Repr. New York: Arno Press.

Hornaday, William Temple. 1889 [2002]. *The extermination of the American bison.* Repr. Washington, DC: Smithsonian Institution Press.

Hunter, Malcolm L. Jr. 1996. *Fundamentals of conservation biology.* Cambridge, MA: Blackwell Science.

Hynes, H. Patricia. 1993. *Taking population out of the equation: Reformulating I=PAT.* North Amherst, MA: Institute on Women and Technology.

Johns, David M. 1999. If there were a cookbook for conservation. Review of *Measures of success* by Richard Margoluis and Nick Salafsky. *Conservation Biology* 13, no. 5: 1228–1229.

Kadet, Anne. 2000. Put your money where our math is. *SmartMoney Magazine* (November 10). http://www.smartmoney.com/mag/index.cfm?story=charitychart.

Kleiman, Devra G., Richard P. Reading, Brian J. Miller, Tim W. Clark, J. Michael Scott, John Robinson, Richard L. Wallace, Robert J. Cabin, and Fred Felleman. 2000. Improving the evaluation of conservation programs. *Conservation Biology* 14, no. 2: 356–365.

Krasner, Stephen D. 1983. Structural causes and regime consequences: Regimes as intervening variables. In *International regimes*, ed. Stephen D. Krasner, 1–21. Ithaca NY: Cornell University Press.

Lipschutz, Ronnie D., and Judith Mayer. 1996. *Global civil society and global environmental governance: The politics of nature from place to planet.* Albany: State University of New York Press.

Margoluis, Richard, and Nick Salafsky. 1998. *Measures of success: Designing, managing, and monitoring conservation and development projects.* Washington, DC: Island Press.

Mattson, David J., and John J. Craighead. 1994. The Yellowstone grizzly bear recovery program: Uncertain information, uncertain policy. In *Endangered species recovery: Finding the lessons, improving the process*, ed. Tim W. Clark, Richard P. Reading, and Alice L. Clarke, 101–129. Washington, DC: Island Press.

Meffe, Gary K., and C. Ronald Carroll, eds. 1997. *Principles of conservation biology.* Sunderland, MA: Sinauer Associates.

Pisanty-Baruch, Irene, Jane Barr, Ed B. Wiken, and David A. Gauthier. 1999. Reporting on North America: Continental connections. *George Wright Forum* 16, no. 2: 22–36.

Ruggiero, Michael A. 1997. Development of biological diversity indicators and targets in the United States. Biological Resources Division, U.S. Geological Survey. Paper presented to the Sixth Global Biodiversity Forum Meeting, U.N.

Headquarters, New York. Exploring Biodiversity Indicators and Targets under the Convention on Biological Diversity, April 3–4.

Salafsky, Nick, and Richard Margoluis. 1999. Threat reduction assessment: A practical and cost-effective approach to evaluating conservation and development projects. *Conservation Biology* 13, no. 4: 830–841.

Shell, Hanna Rose. 2002. Finding the soul in the skin. Introduction to *The extermination of the American bison*, ed. William Temple Hornaday, viii–xxiii. Washington, DC: Smithsonian Institution Press.

Stroud, Richard H. 1985. *National leaders of American conservation*. Washington, DC: Natural Resources Council of America and Smithsonian Institution Press.

Tocqueville, Alexis de. 2000. *Democracy in America*. Chicago: University of Chicago Press.

Turner, Anthony M., Ed B. Wiken, and Nikita Lopoukhine. 1999. Reporting and indicators for protected areas and ecosystems: A national perspective. *George Wright Forum* 16, no. 2: 37–51.

Wapner, Paul. 1996. *Environmental activism and world civic politics*. Albany: State University of New York.

Wilkinson, Todd. 1998. *Science under siege: The politicians' war on nature and truth*. Boulder, CO: Johnson Books.

Wilson, Edward Osborne. 2002. *The future of life*. New York: Alfred A. Knopf.

Wise, Timothy A., Hilda Salazar, and Laura Carlsen, eds. 2003. *Confronting globalization: Economic integration and popular resistance in Mexico*. Bloomfield, CT: Kumarian Press.

Wood, Alexander, Pamela Stedman-Edwards, and Johanna Mang, eds. 2000. *The root causes of biodiversity loss*. London: Earthscan.

World Conservation Monitoring Center. 1992. *Global biodiversity: Status of the earth's living resources*. London: Chapman & Hall.

Y2Y Conservation Initiative. 2003. Kaska, CPAWS sign historic agreement. *Connections* [Y2Y newsletter], no. 6 (Winter), 1–2. http://www.y2y.net/media/connections/winter03web.pdf.

Y2Y Conservation Initiative. 2005. Y2Y Principles for working with aboriginal people. http://www.y2y.net/people/native.asp

Young, Oran R. 1996. Introduction: The effectiveness of international governance systems. In *Global environmental change and international governance*, ed. Oran R. Young, George J. Demko, and Kilaparti Ramakrishna, 1–27. Hanover, NH: University Press of New England.

Young, Oran R., ed. 1999. *The effectiveness of international environmental regimes: Causal connections and behavioral mechanisms*. Cambridge, MA: MIT Press.

Young, Oran R., and Marc A. Levy. 1999. The effectiveness of international environmental regimes. In *The effectiveness of international environmental regimes: Causal connections and behavioral mechanisms*, ed. Oran R. Young, 1–32. Cambridge, MA: MIT Press.

ACKNOWLEDGMENTS

"Most everything I've learned about the desert," wrote Gary Paul Nabhan (1994, 51), "has come from rooting around in the wrong place at the right time." My too short time in the desert—and the mountains—was just the opposite: I seemed to be in the right place at the wrong time. Had I only started my research a few years earlier, I incessantly fumed to myself, I could have had both time and place "right." As it was, I had no choice but to rely on sixty participants in ISDA and Y2Y who *were* there at the right time, to share their memories with me. (See "List of Interviewees.")

And share they did, often sacrificing hours of their already crammed schedules. In talking with them all, I could not help but note that several personas in ISDA were remarkably analogous to those in Y2Y; each initiative had the visionary leaders, the immensely patient stewards, the unassuming and careful analysts, and the relentlessly indefatigable workers. Indeed, the parallels were sometimes so close that I felt like a dazed Dorothy who sees the Kansas farmhands as her traveling companions on the roads of Oz. Though one could never confuse the Sonoran Desert with the Northern Rockies, the good will and the passion of their respective inhabitants were indistinguishable. Not only did they all endure my long interviews, each interviewee read some part of the text—many a chapter or more—and their corrections and suggestions were invaluable. Thus, to my sixty interviewees, I emphatically say thank you.

This book began as a dissertation at the Fletcher School at Tufts University, and I could not have been more fortunate in the choice of my three committee members. First and foremost, I was advised and mentored by the extraordinarily sagacious and remarkably generous Professor William R. Moomaw. Despite his grueling schedule, Bill always thoroughly examined my work with his insightful eye and never failed to give me his undivided attention. He began guiding me through the process of writing this book before I even knew I was going to apply for the Ph.D. program, and I would have given up early on without his unwavering good humor, endless patience, and ability to steer me away from innumerable analytical dead ends. I am also highly indebted to Theodore M. Smith of the Kendall

Foundation, long my stalwart supporter, who has given me the opportunity to keep one foot outside of the cloistered walls of academia. Thank you for your faith in me. Finally, I am grateful to Professor Alan Henrikson, who, in addition to expanding my understanding of U.S. foreign diplomacy in the twentieth century, provided critical assistance at key junctures in my research and writing.

Many thanks go to all the others who have read and commented on individual sections of the book: James Bandler, Bill Broyles, Nina Chambers, John C. Chester Sr., Lael E. H. Chester, Brian Churchill, Mark Cioc, Mark Cleaveland, Steve Cornelius, Juliet Fall, Erica Fleishman, Selma Glasscock, Jim Levitt, Gil Lusk, Bob Peart, Beth Russell-Towe, Nick Salafsky, Paul Schullery, Bill Simoes, Roger Soles, Tracy Taft, and Dorothy Zbicz. Special thanks to David Johns, who read the entire manuscript, and to Tim Clark, who in addition to reading the manuscript planted the first seed of this book in my head. Many thanks also go to Gary Tabor, director of Y2Y Programs at the Wilburforce Foundation, who found time to read several draft chapters. Although I have been studying environmental issues since an undergraduate, my real education in conservation began during a road trip with Gary from Boston to the 1996 Montreal IUCN World Conservation Congress. In twelve short hours, he taught me (mostly by example) what it means to be a passionate and effective conservationist. Finally, many apologies in advance to whomever I have neglected to thank for commenting on my drafts.

Heartfelt thanks go to Hector Laguette-Rey, who transcribed and translated the interviews that I conducted in exceptionally mediocre Spanish—although any and all factual errors, inconsistencies, and misrepresentations throughout this book are entirely his fault. In all seriousness, I could offer no better proviso for this book than to quote Canadian novelist Margaret Atwood (1972), who wrote about one of her own books that the "accuracies and fine points in this book were for the most part contributed by others; the sloppy generalizations are my own."

My editor at Island Press, Barbara Dean, deserves a medal for putting up with me, my delays, and my petulancies throughout the process of getting this book completed. Many a psychopathic rampage through the streets of Central Square, Cambridge, was averted by her constant encouragement. Laura Carrithers and Cecilia González were also exceptionally helpful in the final stages of preparing the book.

Thanking your family is sort of like telling a friend that you are sorry about burning down his house; it needs to be done but feels woefully inadequate. Nonetheless, my entire family deserves credit for not wringing my neck while I labored over the book. I trace my love of reading to my mother, Clara Mills Chester (1931–1985), while I trace my love of writing to my father, John C. Chester Sr. (a former speechwriter in the House of Representatives, who taught me how to write despite my belief in the ninth grade that I had better things to do than compose a paper on the Alaska pipeline). Compared to everything they have given me, this book is but a pamphlet. Special thanks to my aunt and uncle, Marion and Verne Read, who have been my adopted parents for many years. Thanks to my role-model brother John C. Chester Jr., whose *Ares: The God of War in Greek Epic, Lyric, and Tragic Poetry*, a thesis composed before personal computers with the ancient Greek hand-lettered in after typing, sat on my desk as an inspiration (and constant reminder of how easy I have it). Thanks to my sister, Isabelle P. Chester, whose creativity and support have taught me that true understanding of anything of importance requires both the head and the heart. And thanks to their families and to my stellar in-laws, the Hiam-McAlpine clan—all of whom have given me more than I ever dared hope.

"Earth's the right place for love," wrote Robert Frost, and I'm grateful to my wife, Lael, and sons, Sam and Caleb, for continually reminding me of that. This book is most fondly dedicated to them.

References

Atwood, Margaret Eleanor. 1972. *Survival: A thematic guide to Canadian literature.* Toronto, Canada: Anansi.

Nabhan, Gary Paul. 1994. *Desert legends: Re-storying the Sonoran borderlands.* New York: Henry Holt.

LIST OF INTERVIEWEES

Peter Aengst	Wendy Francis	Luther Propst
Rob Ament	Manuel Gonzalez	Tony Ramon
Dave Augeri	Larry Gray	Ray Rasker
Mike Bader	Joseph Joaquin	Bart Robinson
Ken Barrett	Wendy Laird	Wayne Sawchuk
John Bergenske	Stephen Legault	Mike Sawyer
Peggy Turk Boyer	Anne Lemorande	Bob Schumacher
Bill Broyles	Anne Levesque	Chuck Schwartz
Alberto Búrquez	Ed Lewis	Michael Scott
Reggie Cantú	Harvey Locke	Tom Skeele
Dominic Cardea	Marcy Mahr	Ceal Smith
Kenia Casteñeda	David Mattson	Craig Stewart
Barb Cestero	Beau McClure	Gary Tabor
Steve Cornelius	Joaquin Murrieta	Christine Torgrimson
Lance Craighead	Gary Paul Nabhan	Sue Tout
Paul Crawford	Carlos Nagel	Kevin Van Tighem
Katie Deuel	David Ortiz Reyna	Louisa Willcox
Lorraine M. Eiler	Brian Peck	Carlos Yruretagoyena
Bob Ekey	Juri Peepre	Julie Zammuto
Exequiel Ezcurra	Jim Posewitz	Rick Zammuto

LIST OF ABBREVIATIONS

ALSA	American Lands Sovereignty Act
A2A	Algonquin to Adirondacks Conservation Initiative
AWA	Alberta Wilderness Association
AWL	American Wildlands
AWR	Alliance for the Wild Rockies
BECC	Border Environmental Cooperation Commission
BLM	Bureau of Land Management (U.S.)
BMGR	Barry M. Goldwater Range
BR	biosphere reserve
BSP	Biodiversity Support Program
B2B	Baja to Bering Sea Marine Reserve Initiative
CBC	community-based conservation
CBD	Convention on Biological Diversity
CC	Coordinating Committee (Y2Y)
CCA	Carnivore Conservation Area
CCE	Crown of the Continent Ecosystem
CCI	Comité Consenso Internacional
CCR	Comité Consenso Regional
CEDO	Centro Intercultural de Estudios de Desiertos y Océanos
CES	Chetwynd Environmental Society
CITES	Convention on International Trade in Endangered Species
CMS	Convention on Migratory Species
CPAWS	Canadian Parks and Wilderness Society
CPNWR	Cabeza Prieta National Wildlife Refuge
CSI	Consenso Sonorense Internacional
DOD	Department of Defense (U.S.)
DOI	Department of the Interior (U.S.)
ECOSOC	Economic and Social Council (UN)
EKES	East Kootenay Environmental Society
ESA	Endangered Species Act
FPN	Friends of ProNatura
FWS	Fish and Wildlife Service (U.S.)
GBF	Great Bear Foundation
GEF	Global Environment Facility
GIS	geographic information system
GNP	Glacier National Park
GYC	Greater Yellowstone Coalition
GYE	Greater Yellowstone Ecosystem
H2O	Highlands to Ocean Initiative
HUBEC	Headwaters Unfragmented Biodiversity Ecosystem Coalition
IBP	International Biological Program
ICC	International Coordinating Council (MAB)

LIST OF ABBREVIATIONS

ICDP	Integrated Conservation and Development Project
IER	international environmental regime
IGY	International Geophysical Year
IJC	International Joint Commission
IPY	International Polar Year
ISDA	International Sonoran Desert Alliance
ITTO	International Tropical Timber Organization
IUCN	World Conservation Union (International Union for the Conservation of Nature and Natural Resources)
LOI	Letter of Intent
LRMP	Land and Resource Management Plan
MAB	Man and the Biosphere
MKMA	Muskwa-Kechika Management Area
MOU	Memorandum of Understanding
NAFTA	North American Free Trade Agreement
NCDE	Northern Continental Divide Ecosystem
NGO	nongovernmental organization
NPS	National Park Service (U.S.)
NREPA	Northern Rockies Ecosystem Protection Act
NUBEC	Northern Unfragmented Biodiversity Ecosystem Coalition
NWR	National Wildlife Refuge
ORPI	Organ Pipe Cactus National Monument
PRI	Institutional Revolutionary Party (Mexico)
ProNatura	Asociación Mexicana Pro-conservacion de la Naturaleza
RMEC	Rocky Mountain Ecosystem Coalition
SAHR	Ministry of Agriculture and Water Resources (Mexico)
SDEP	Sonoran Desert Ecosystem Partnership
SDNP	Sonoran Desert National Park
SEDUE	Ministry of Urban Development and Ecology (Mexico: Secretaría de Desarollo Urbano y Ecología
SI	Sonoran Institute
STCL	Save-the-Cedar League
TBPA	transboundary protected area
TFCA	transfrontier conservation area
TGB	Trail of the Great Bear
TNC	The Nature Conservancy
TON	Tohono O'odham Nation
TWP	The Wildlands Project
UBEC	Unfragmented Biodiversity Ecosystem Coalition
UNEP	United Nations Environmental Programme
UNESCO	United Nations Educational, Scientific, and Cultural Organization
USDA	U.S. Department of Agriculture
USFS	U.S. Forest Service
WCPA	World Commission on Protected Areas
WPCCC	Western Pima County Community Council
WREF	Wild Rockies Earth First!
WWF	World Wildlife Fund
YNP	Yellowstone National Park
Y2Y	Yellowstone to Yukon

A NOTE ON WEB NOTES

The author and the editors at Island Press have decided to do something a bit different with the notes to this book. Rather than weigh the book down with many pages of notes—something that would have both raised the price of the book and, for the vast majority of readers, constituted an unconscionable waste of natural resources—we are making the notes available at http://www.islandpress.org/conservation_across_borders/webnotes.pdf. These notes—or what we now call "Web notes"—largely provide further source materials, background information, and anecdotal narratives. If (author's note: the editors will not let me say "when") inaccuracies are discovered in the book, notice will be given in these Web notes. While this information is mainly directed toward researchers, other readers may find it useful to print these out and use them as a somewhat hefty bookmark.

INDEX

A

Abbey, Edward, 62
Abbott, Carl, 76
Abitibi Consolidated Inc., 198
Action Plan for Biosphere Reserves of 1984, 42
Adams, W. M., 36, 41
Adirondack-Champlain Biosphere Reserve, 111
Aengst, Peter, 158, 162, 185, 188, 196
African Convention on the Conservation of Nature and Natural Resources of 1968, 22
Aguirre, A. Alonso, 8
Ajo School Arts Partnership, 120
Ajo Valley, 70, 72–73, 78, 98, 112, 119–21
Akeley, Carl, 22
Alaimo, Carol Ann, 76
Alaska Maritime National Wildlife Refuge, 19
Alaska-Yukon-British Columbia World Heritage Area, 14
Alberta Environment, 141
Alberta National Park, 22
Alberta Wilderness Association, 157, 165
Alexander, Ben, 163–64
Algonquin to Adirondacks Conservation Initiative (A2A), 206
Allelic diversity, 6
Allen, Paula Gunn, 77
Alliance for the Wild Rockies (AWR), 148, 151, 152, 167, 192
Alto Golfo de California y Delta del Rio Colorado Biosphere Reserve, 43, 57
Amargos Complex, 66

Ament, Rob, 166, 184, 187, 190, 193
American Lands Sovereignty Act (ALSA) of 1997, 112
American militia movement, 110
American Wildlands, 147, 166, 184, 188, 193
Amigos de Pinacate, 64
Anasazi people, 66
Ancestral (Ancient) Puebloans, 66
Anderson, Susan, 71, 74, 78, 81, 82, 93
Andrian, Giorgio, 37
Archer, Fiona, 28
Arizona Desert Wilderness Act of 1990, 98
Arizona Game and Fish Department, 83
Arizona-Sonora Desert Museum, 99, 106, 113
Arizona-Sonora Memorandum of Understanding (MOU), 112–16
Asociación Mexicana Proconservacion de la Naturaleza, 64
Aunger, Robert, 203

B

Babbitt, Bruce, 91, 114
Bader, Mike, 148, 152, 192
Bailey, Robert G., 34, 44
Baja to Bering Sea Marine Reserve Initiative (B2B), 206
Banff National Park, 142, 150, 166, 179–80, 183, 200
Barajas, Maria Elena, 93, 113
Barcott, Bruce, 181
Barios, Joseph, 76
Barry M. Goldwater Range (BMGR), 57, 119

Batisse, Michel, 34–37, 40, 41, 42, 60, 105
Bears, 135–36, 151, 218, 223
"Bears of Banff Inbreeding Game," 163
Beck, Warren A., 56
Beltrones, Manlio Fabio, 112, 113
Bengeyfield, Pete, 168
Bennett, Bruce, 172, 207
Bennett, Peter S., 68
Benvenisti, Eyal, 15, 18
Berg, Peter, 34
Bergenske, John, 171
Berkman, Paul Arthur, 35
Bernstein, Steven F., 195
Besançon, Charles, 16, 22
Bienen, Leslie, 194
Big Bend National Park, 115, 149
Binational Network of Sonoran Desert Biosphere Reserves, 112–13
Biodiversity, 3–8
  academic disciplines and, 8
  adoption of concept of, 6
  defined, 6
  global extinction and, 5–7
  homo sapiens and, 5, 7
  indicators of, 222–23
  in North America, 4–5
  root causes of loss of, 221–23
  threatened species, 5–6
  transborder definition and, 15–17
  Y2Y and, 153–55
Biodiversity Convention of 1992, 7, 16, 31–32, 110, 222
Biodiversity Support Program (BSP), 225

252 INDEX

Biodiversity (Wilson), 6, 7
Bioregions, 33–44
"Biosphere Conference" of 1968, 35, 36, 37
Biosphere reserves, 37–43
   Arizona-Sonora MOU, 112–16
   defined, 38
   demise of international efforts for, 108–12
   development of, 37–39
   expanding, in the Western Sonoran desert, 97–107
   fundamental goals of, 38–39
   land-use patterns and Y2Y, 39–42
   local participation and ISDA, 42–43
   in Mexico, 59–61
   as political pariah, 107–18
   transborder cooperation without, 116–18
   U.S.-Mexico LOI, 114–16
   see also specific reserves
Biosphere Reserves in Action: Case Studies of the American Experience (Condo and Bishop), 99, 102–103
Birnie, Patricia W., 18
Bishop, Sarah H., 60, 64, 65, 70, 85, 99, 102, 107
"Black helicopters," 110–11, 112, 175
Blake, Tupper Ansel, 64, 92
Blue Ribbon Coalition, 174
Bob Marshall Wilderness Area, 142, 149
Bonn convention of 1979, 16, 18
Border Environmental Cooperation Commission (BECC), 91
Border Environmental Education Group, 91
Border Grizzly Project, 151
Border Land Use Forum of 1992, 69–75, 78, 81, 84, 86, 93, 94, 97, 99, 119, 231
Borders, 1–3
Borealis, 144, 145, 146
Botkin, Daniel B., 136

Boundary Waters Treaty of 1909, 19
Bowden, Charles, 67
Bow Valley, 183
Boyd, David R., 137, 141
Boyd, Diane, 146, 148
Boyle, Alan E., 18
Boy Scouts of America, 56
Bozeman Chronicle, 175
Bridge of the Americas, 2
Bridging Borders Conference of 1994, 86–87, 97, 99
British Columbia Ministry of Forests, 141
British Columbia Parks, 141
Brosius, J. Peter, 29
Brown, David E., 56, 57
Broyles, Bill, 54, 72, 73, 119
Brunnée, Jutta, 14
Brunner, Robert, 24, 30
Burke, David, 164
Búrquez, Alberto, 63, 64, 65, 87, 92, 94
Bush, George H. W., 7

C

Cabeza de Vaca, Álvar Núñez, 1, 54
Cabeza Prieta Game Range, 57
Cabeza Prieta National Wildlife Reserve, 61, 62, 71, 83, 89, 98, 103, 113, 119
Calgary Herald, 166
Camp, Martha, 104
Camp Fire Club of America, 219, 235
Camp-fires in Desert and Lava (Hornaday), 217–18
Camp-fires in the Canadian Rockies (Hornaday), 217–18
Canadian Broadcasting Corporation, 164
Canadian Parks and Wilderness Society (CPAWS), 143, 144, 147, 148, 149, 152, 157, 158, 197–203, 206
Cañon de Santa Elena Protected Area, 115
Cantú, Reggie, 78, 111, 113
Carabias, Julia, 114
Cardea, Dominic, 111
Carlton, James T., 6

Carmony, Neil B., 56, 57
Carnivore Conservation Areas (CCAs), 146–47
Carroll, C. Ronald, 220
Carroll, Francis M., 2
Castillo-Sánchez, Carlos, 63, 87, 92, 94
Castle Crown Wilderness, 183
Center for Marine Conservation, 206
Centro Ecológico de Sonora, 63, 65, 83, 92, 93, 95
Centro Intercultural de Estudios de Desiertos y Océanos (CEDO), 92, 95
Cestero, Barb, 171, 199
Chadwick, Douglas H., 171
Chape, Stuart P., 16, 39
Chenoweth, Helen, 117
Chester, Charles C., 43, 115, 119, 149, 153, 164, 171, 174, 185, 195, 227
Chetwynd Environmental Society (CES), 152, 189, 197
Cheviot coal mine, 183
Christen, Kris, 7
Christensen, John R., 83, 224
Churchill, Brian, 198–99
Cisneros, Jose A., 114, 115
Civil society, 4
   government vs., 234–38
Clark, Ella Elizabeth, 141
Clark, Tim, xiii, 152, 193, 204
Clarke, Robin, 223
Clinton, Bill, 7, 112, 114, 119
Coate, Roger A., 59
Cocopah people, 104
Colosio, Luis Donaldo: assassination of, 96
   Pinacate Biosphere Reserve and, 91–96
Comite Consenso Internacional (CCI), 53, 79, 81, 85, 88, 98
Comite Consenso Regional (CCR), 81, 85, 91
Commission on Environmental Law, 32
"Completion of the Sonoran Desert Biosphere Network along the U.S./Mexico Border," 104–105

INDEX 253

Comus, Patricia Wentworth, 62
Conca, Ken, 18
Condo, Antoinette J., 60, 64, 65, 70, 85, 99, 102, 103, 107
Conference of the Parties, 32
Connections, 163
Connectivity principle, 146
Consenso Sonorense Internacional (CSI), 80, 81
Conservation and Rational Use of the Environment, 35
Conservation biology, 145–46, 147, 193–97
Conservation Biology, 193
Conservation effectiveness in the territories of change, 217–38
  civil society vs. government, 234–38
  communication systems, 231
  comparative approach to, 228–34
  complexity in measuring, 220–24
  constituency inclusion, 230–31
  dependent variables and, 229
  evaluation and monitoring (E&M), 225
  formulas for, 220–22
  historical methodology, 228–29
  independent variables and, 229–34
  international regime models, 226–28
  landscape vision, 232–33
  leadership dynamics, 232
  mission breadth, 230
  multiple meanings of, 219–24
  organizational assessments, 225–26
  political backlash, 232
  political science methodology, 228–29
  project and program oriented evaluations, 224–25
  scientific participation and support, 231
  threat reduction assessment, 225
Conservation Foundation, 36
Conservation Measures Partnership, 225
Conservation Strategy for Large Carnivores in Canada, A, 146
Conservation successes of Y2Y, 182–203
  conservation "processes," 194
  electronic mailing list (EML), 185–87, 231
  increased funding, 191–93
  inspiration in a "landscape of hope," 187–88
  learning across borders, 188–90
  Muskwa-Kechika, 197–200
  networking, 183–87
  peacemaking among alpha conservationists, 190–91
  science and conservation planning, 193–97
  summary of, 182–83
  Yukon, 200–203
"Conservation Terrestrial," 35
Convention for the Preservation of Wild Animals, Birds and Fish in Africa of 1900, 21–22
Convention for the Prevention and Protection of Fur Seals of 1911, 19
Convention of International Trade in Endangered Species (CITES) of 1973, 16
Convention on Biological Diversity (CBD) of 1992, 7, 16, 31–32, 110, 222
Convention on Conservation of Migratory Species of Wild Animals (CMS) of 1979, 16, 18
Corkeron, Peter J., 5
Cornelius, Steve, 111–16, 118
Corps of Discovery, 134–36
Corridors of Life project, 147
Council of Europe, 29
Craighead, Frank, 168
Craighead, John, 168, 223
Craighead, Lance, 172
Crown of the Continent Ecosystem (CCE), 148–49, 167, 169–70
Crow's Nest Pass, 183
Curley School complex, 119–20

D

Dabelko, Geoffrey D., 18
Daily, C. G., 7
Dasmann, Raymond F., 20, 34, 35–36, 42
Davis, Wade, 8
Dawkins, Richard, 203
Dawson, Chad P., 142
Death Valley National Park, xiii
DeConcini, Dennis, 84, 85, 86
Deh Cho First Nation, 202, 203
de Klemm, Cyrille, 34, 40, 60, 111
Democracy in America (Tocqueville), 237–38
Dena Tuna, 173
DeSombre, Elizabeth R., 4
Deuel, Katie, 162
Diamond, Donald, 70
di Castri, Francesco, 37–38, 42
Dorsey, Kurkpatrick, 4, 18, 20, 217
Douglas, Ross, 24
Draper, Malcolm, 28
Drug trafficking, 75–76
Dusard, Jay, 2, 149
Dutton, Shelia, 28
Dyer, M. I., 40

E

Earth First!, 62, 151, 152
Earth Summit of 1992, 7
Ecological Balance and Environmental Protection Act of 1988, 60
Ecological processes principle, 146
Ecology of North America, The (Shelford), 139

Ecoregionalism, 34
Ecotourism, 149–50
Ehrlich, Anne H., 221
Ehrlich, Paul, 7, 221, 222
Eiler, Lorraine, 67, 68, 72–73, 78, 79, 96, 97–98
Eisner, Thomas, 8
Ekey, Bob, 157, 171, 192
Electronic communications, 185–87, 231
Ellis, Cathy, 164
El Rosario sanctuary, 15
Emanuel, Robert, 66
Endangered Species Act (ESA), 139, 140, 180–181
Environmental Conservation (Dasmann), 34
Environmental peacekeeping, 18
Environmental security, 18
Environment Committee of the Arizona-Mexico Commission, 43
Equinox, 158
Erickson, Winston P., 66
EUROPARC Federation, 31
European Community (EC), 16
European Outline Convention on Transfrontier Co-operation between Territorial Communities or Authorities of 1980, 29
Evangelical Environmental Network, 180–81
Explore Magazine, 164
Exploring Common Ground on the Dena Tuna: . . . , 173
Ezcurra, Exequiel, 37–38, 40, 42, 59, 62–65, 71, 110
Pinacate Biosphere Reserve and, 92–97

F

Fall, Juliet J., 37
Felger, Richard Stephen, 56, 57, 58, 62, 87, 119
Fifth Discipline, The (Senge), 178–79
Finkel, Michael, 164
First Nations (Native Americans), 141, 173, 197, 202–203, 230–31
Fisher, Richard D., 61
Fitter, Richard, 22
Flannery, Tim F., 5
Fleming, Theodore H., 120
Fletcher, Susan R., 111
Flinton, Fiona, 29
Flores, Floyd, 83, 105
Ford Foundation, 119
Foreman, David, 62, 87, 147, 152, 233
Fort Nelson LRMP, 198, 199
Fort St. John LRMP, 198
Fossey, Diane, 22
Foundations of Success (FoS), 225
Francis, Wendy, 156, 157, 159, 173, 184–85, 189
Franklin, Jerry F., 34, 37–40
French Forest Ordinance of 1669, 20
Friends of ProNatura (FPN), 64, 68, 69, 88, 89, 100
ISDA and, 79–80
Friends of the Earth, 236
Frontera, La, 76–78

G

Gadd, Ben, 139
Gadsden Purchase, 2, 58
Gaillard, David L., 204
Gailus, Jeff, 99
Gamboa, Jorge Luis, 87
Garay, Ortiz, 90
García Blanco, Marco Antonio, 82
Gartlan, Steve, 23, 30
Gauthier, David A., 223
Gawthrop, Daniel, 2
Gay, A. E., 73
Gell-Mann, Murray, 8
Genetic diversity, 6
Geographic information systems (GIS), 94
Gilbert, Vernon C., 58
Glacier Biosphere Reserve, 43
Glacier National Park, 21, 142, 148–49, 169
Gladwell, Malcolm, 165
Global Biodiversity Forum, 222
Global Environment Facility (GEF), 33
Global Information System (GIS), 88
Global Partnership for Peace Parks, 32
Global Transboundary Protected Area Network (GTPAN), 33
Goedeke, Theresa L., 38, 110, 111
Goetel, Walery, 21
Golley, Frank B., 35, 36–37
Gómez-Pompa, Arturo, 60, 104
Google, 185
Gran Desierto de Altar, 61–62
Grand Teton National Park, 142
Grant, Marie, 150
Grant, Ulysses S., 19
Gray, Larry, 201
Great Bear Foundation, 149
Greater Ecosystem Alliance, 144, 148
Greater Yellowstone Coalition, 152, 161, 169, 189
Greater Yellowstone Ecosystem (GYE), 168–69, 223
Great Limpopo Transfrontier Park, 44
Greenpeace, 236
Gregg, William P., Jr., 36, 41, 54, 60, 61, 69, 70–71
Griffin, John, 15–16, 23
Grifo, Francesca, 8
Grinnell, George Bird, 148
Grumbine, R. E., 58
Guia Roji, S. A. de C. V, 56
Gulf of Maine Council for the Marine Environment, 14

H

Haas, Peter, 195
Haase, Ynez D., 56
Haberl, Helmut, 5
Haines, Aubrey L., 142
Halffter, Gonzalo, 42, 59–61, 64
Hamilton, Lawrence S., 24–27, 30
Hammill, Anne, 22
Hansard, 172
Harris, Elizabeth, 17
Harrison, J. M., 24, 37
Hartmann, Bill, 87
Harvard Business Review, 226
Harvey, Ann, 193
Harvey, Philip D., 226

INDEX 255

Hayden, Julian D., 61–66
Hayes, Derek, 135
Hayes, Peter, 23
Headwaters UBEC, 150–51
Heijnsbergen, P. van, 8, 18, 21
Helvarg, David, 59, 111
Hendee, John C., 142
Henderson, John Brooks, 19
Herman, Otto, 18
Herrero, Stephen, 193
Heuer, Karsten, 146, 147, 164, 172
Heywood, Vernon H., 223
Hia Ced O'odham Nation, 61, 65–68, 72–73, 94, 230
Highlands to Ocean Initiative (H2O), 206
Hiss, Tony, 206
Hohokam people, 66
Holdgate, Martin, 35, 36
Holdren, John P., 222
Holland, M. M., 40
Homer-Dixon, Thomas F., 18
Hoover, Herbert, 103
Hornaday, William T., 217–19, 234, 235
Hough, John, 39, 40, 41
Hughes, Jennifer B., 7
Hughes, Ross, 29
Human rights, 33
Hummel, Monte, 146, 147, 148
Hunter, Malcolm L., Jr., 220
Hutton, Carrie, 164
Hynes, H. Patricia, 224

I

Illegal immigration, 75–76
Indian Village, 78
Indigenous cultures:
   ISDA and, *see* O'odham people
   Y2Y and, 141, 173, 198, 202–203, 230–31
Institutional Revolutionary Party (PRI), 92, 93, 96
Instituto Nacional de Ecología, 60, 64, 71
Integrated conservation and development projects (ICDPs), 28–29
Inter-Hemisphere Resource Center, 112–13, 115
International Biological Program (IBP), 35–37

International borders, 1–3
   Canada-U.S., 14, 75–78
   Mexico-U.S., 14, 109–12
International Conference of Experts on a Scientific Basis for Rational Use and Conservaton of the Biosphere of 1968, 34
International conservation, 3–4
International Convention for the Protection of Birds Useful to Agriculture of 1902, 18
International Coordinating Council (ICC), 36, 37, 38
International Council of Scientific Unions, 35, 37
International Covenant of Environment and Development of 1994, 32
International environmental regimes (IERs), 226–28
International Geophysical Year (IGY), 35
International Joint Commission (IJC), 19
International Polar Year (IPY) of 1882, 35
International relations, 234–38
International Sonoran Desert Alliance (ISDA), xiii–xv, 34, 53–133, 174
   agreements to agree, 112–16
   biosphere reserve as political pariah, 107–18
   border culture and, 76–78
   Border Land Use Forum of 1992, 69–75
   civil society and, 4
   conceptual foundations of, 4
   conservation effectiveness of, *see* Conservation effectiveness in the territories of change
   conservation efforts not participated in by, 118–19
   contemporary land management patterns, 54–58

   cooperation without the biosphere label, 116–18
   defining a mission, 79–81
   designation of the Pinacate Biosphere Reserve, 91–97
   diversity on the board of, 82–84
   drug trafficking and, 75–76
   early humans and, 65–66
   early programmatic agenda, *see* Programmatic agenda of ISDA, early
   economic slump of 1984 and, 78
   effects of, 118
   expanding biosphere reserves and, 97–107
   faces of, 73–75
   finding a name, 78–79
   goals of, 79
   illegal immigration and, 75–76
   incorporation debate, 81–82
   indigenous cultures and, 94–97
   MAB in Mexico, 59–61
   MAB in the U.S., 58–59
   map of, 55
   Mexican government and, 91–97
   as new form of transborder conservation, 43–44
   O'odham people and, 65–68
   Pinacate region and, 61–68
   principle threads of, 75
   region covered by, 53–54
   SI-ORPI Cooperative Agreement and, 84–85
   Sonoran Institute and, 69–75
   today and tomorrow, 118–21
   as voice of and for the desert, 75–84

256 INDEX

International Tropical Timber Organization (ITTO), 32, 33
International watercourses, 18
Internet, 185–87
Ishmael (Quinn), 180
IUCN World Commission on Protected Areas, 171

J
Jacobson, Susan Kay, 8
Jarrell, Randall, 34
Jasper National Park, 142, 143, 149, 150, 167, 200
Jefferson, Thomas, 135
Jeffries, Mike J., 6
Jessen, Sabine, 206
Joaquin, Joe, 95
Johns, David, 142, 144, 147, 148, 153, 157, 180–81, 204, 225
Johnson, Lyndon B., 103
Jonkel, Charles, 149–50, 151
Joquin, Angelo, Sr., 61, 67, 68
Julian Hayden Award, 97
Juntos program, 88–89

K
Kahn, Tamar, 28
Kananaskis Country, 182–83
    Y2Y meetings held in, 144, 148, 153, 155, 158
Kaska Dene First Nation, 203
Kaus, Andrea, 60
Keiller, Douglas, 80, 86
Kendall Foundation, 159
Keystone National Policy Dialogue on Ecosystem Management, 109
Kingsolver, Barbara, 78
Kino, Father Eusebio, 54
Kitt Peak National Observatory, 89
Kleiman, Devra G., 224, 225
Kolbe, Jim, 84
Kootenay National Park, 142
Krakow Protocol of 1924, 20–21
Krasner, Stephen D., 226
Krech, Shepard, 66
Krutilla, John V., 6
Kukura, Rudolf, 21

L
Lagault, Stephen, 143, 157, 165, 167, 172, 182–83, 187, 192, 199
Laguna Madre Binational Initiative on the Gulf of Mexico, 14
Laird, Wendy, 60, 61, 62, 65, 68–117
Lamb, Henry, 112
Land and Resource Management Plans (LRMPs), 197–99
Landjouw, A., 22
Land Use Changes in the Western Sonoran Desert Border Area (Land Use Forum of 1992), 69–75, 78, 81, 84, 86, 93, 94, 97, 99, 119, 231
Large-landscape approach to conservation, 147–53
LaRue, Steve, 91, 94
La Ruta de Sonora, 90
LaSalle Adams Foundation, 192
Latin American Zoological Conference, 60
Law of the Yukon, The (Service), 176, 177
League of Nations, 22
LeDuff, Charlie, 76
Lee, David W., 20
Lemonrande, Anne, 158
Leopold, Luna, 36
Lerch, Natalie, 206
Levesque, Anne, 157, 172, 195–96, 199
Levin, Donald A., 5
Levin, Phillip S., 5
Levy, Marc A., 226
Lewis, Ed, 170–71, 184, 188, 196
Lewis and Clark, 134–36, 139
Lieberman, Susan, 76
Liidlii Kue First Nation, 202
Lincoln Institute of Land Policy (LILP), 71–72
Listo, Sylvester, 64, 91
Locke, Harvey, 137, 142–49, 152, 153, 155, 161, 177–83, 196, 199, 201, 202

London Convention Relative to the Preservation of Fauna and Flora in their Natural State of 1933, 21, 22, 23
Lopez, Danny, 87
Los Angeles Times, 171
Louka, Elli, 26, 33
Lovejoy, T. E., 6
Low, Mike, 198
Lowey, Mark, 142, 164
Luke Air Force Base, 83
Luke Gunnery Range, 57
Lusk, Gil, 149, 169
Lyster, Simon, 18, 22

M
MacArthur Fellowship, 99
McClure, Beau, 53
McCoy, Michael, 139
MacDonald, Graham A., 21
Mace, Richard D., 149
McGuire, Randall H., 67
McHugh, Lois, 110
Mackenzie, Alexander, 135, 139
Mackinnon, Kathy, 32
McManus, Roger E., 6
McMillon, Scott, 164
McNeely, Jeffrey A., 8
Maderas del Carmen Protected Area, 115
Maestros y Niños del Desierto, 88–89
Magin, Chris, 39
Magurran, Anne E., 8
Mahr, Marcy, 176–77, 194, 196
Man and the Biosphere Program (MAB), 34–43, 92
    development of, 37–39
    international, 111
    land-use patterns and Y2Y, 39–42
    local participation and ISDA, 42–43, 69, 74, 98
    in Mexico, 59–61
    origins of, 34–37
    Reserve Committee, 113
    as unnecessary, 117–18
    in the U.S., 58–59, 99–17, 117–18
    see also Biosphere reserves

INDEX 257

Manes, Christopher, 62
Mapimi Biosphere Reserve, 60
Margoluis, Richard, 224, 225
Marynowski, Susan, 147
Mattson, David, 142, 206, 223
Mauro, Melissa, 194, 196
Mayhood, Dave, 140
Medeiros, Paul, 147
Meffe, Gary K., 220
Meier, Christopher, 206
Memes, 203–204
Mercer, Rick, 134
Merchant, Carolyn, 19
Merrill, Troy, 142, 156
MexAmerica, 76–78
Mexican Ministry of Agriculture and Water Resources (SAHR), 63, 64–65
Mexican Ministry of Environment and Natural Resources, 115
Mexican Ministry of Urban Development and Ecology (SEDUE), 64–65
Mexico City, 94
Meyer, Judith, 177
Michilfa Biosphere Reserve, 60
Migratory Bird Treaty of 1916, 19–20
Migratory species, 18–20
Miistakis Institute, 156
Milich, Lenard, 18
Minera Hecla, 90
Mitchell, Alanna, 142, 155
Mitchell-Banks, Paul J., 197, 198
"Model Agreement on the Creation and Management of Transfrontier Parks," 29
Moore, Josiah, 72
Mount Revelstroke National Park, 142
Mulongoy, Kalemane Jo, 16
Mumford, Lewis, 33
Munro, Guillermo, 87
Muro, Mark, 113
Murrieta-Saldwar, Joaquin, 43, 59
Muskwa-Kechika Management Area (MKMA), 143, 152, 183, 197–200
Myers, Norman, 7

N

Nabhan, Gary Paul, 59, 64, 66, 67, 71, 87, 99, 100, 104–107, 118
Nagel, Carlos, 64, 68–88, 97–105, 109, 113, 118
Nahanni National Park, 142
National Big Horn Sheep Center, 91
National Geographic, 168
National Geographic Society, 171
National Park Foundation, 89
National Project on the Protected Natural Areas of Mexico, 104
National Public Radio, 164
National Science Foundation, 58
National Wildlife Refuge (NWR), 57
Natural Resources Defense Council, 152
Nature Conservancy, The (TNC), 63, 71, 92, 93, 95, 105, 225
Nature of Things, The, 171
Naylor, Valarie J., 114, 115
"Nectar corridors," 120
New Challenge, The, 163–64
New Cornelia copper mine, 78
Newman, Peter C., 135, 147
Newquist, Susan, 142, 146, 164, 172
New York Times, The, 163
New York Zoological Society, 219, 235
Nimkin, David A., 107
Nongovernmental organizations (NGOs), 4, 25, 84, 113, 147, 225, 235–36
Non-numerical, Unstructured Data: Indexing, Searching, and Theorizing (NUD*IST), xiv
Nonprofit organizations, 225–26
Norse, Elliot A., 6
North American Free Trade Agreement (NAFTA), 72, 93, 94
Northern continental divide ecosystem (NCDE), 149
"Northern Forest," 2
Northern fur seal, 19
Northern Rockies Conservation Cooperative, 152, 193
Northern Rockies Ecosystem Protection Act (NREPA), 148, 152, 168
Northern UBEC, 150–51
Noss, Reed F., 7, 147

O

Ocaña, Dr. Samuel, 63–64, 93
O'odham people, 54, 71, 78, 232
   see also Hia Ced O'odham Nation; Tohono O'odham Nation (TON)
Oppenheimer, Andres, 76, 77, 92
Organization for Economic Cooperation and Development (OECD), 16
Organ Pipe Biosphere Reserve, 43
   designation of, 58–59
Organ Pipe Cactus National Monument (ORPI), 56, 58–59, 62, 64, 68, 71, 83, 89, 103, 104, 106, 109, 113, 119, 120
   Cooperative Agreement with SI, 85, 101
Ozark Man and the Biosphere Reserve, 111

P

Page, Jack, 65
"Panel on the Ecological Integrity of Canada's National Parks," 172
Pannell, Kerr & Foster, 150
Papago people, 66, 67, 72
Papaguería, La, 54
Paquet, Paul, 146
Parezo, Nancy J., 54, 66
Parfit, Michael, 62
Parks Canada, 141, 160, 164, 172
Parque Natural del Pinacate y del Gran Desierto, 71
Parra-Salazar, Ivan E., 115
Pastor, Ed, 84
Pavlakovich-Kochi, Vera, 76
Payne, Roger, 5

INDEX

Peace parks, 20–33, 119, 142
  other names for, 23
  as public face of transborder conservation, 23–26
  see also Transboundary protected areas (TBPA)
Peace Parks Council, 32
Peace Parks Foundation, 32
Pearson, Gina, 56, 57, 59, 61–65, 88, 89, 112, 114, 116
Peck, Brian, 167, 169–70, 184, 190
Peepre, Juri, 200–203
Peine, Mary Anne, 151–52
Pelican Island National Wildlife Refuge, 19
Pew Scholarship for Conservation and the Environment, 99
Phelps-Dodge, 78
Phillips, Steven J., 62, 196
Phoenix, Arizona, 76–77
Pieniny National Park, 21
Pima County Planning Division, 83
Pimería Alta, La, 54
Pinacate region, 61–68
Pinacate y Gran Desierto de Altar Biosphere Reserve, 43, 57, 89, 109, 115
  designation of, 91–97
Pisanty-Baruch, Irene, 223
Planet Drum, 34
Platt, Rutherford H., 8
Pluie (wolf), 146, 163
Polzer, Charles W., 54
Poore, Duncan, 32
Portland State University, 157
Portman, Karla, 98
Posewitz, Jim, 174
Predator Conservation Alliance, 166, 188
Preston, William, 59
Princeton University, 134
Programmatic agenda of ISDA, early, 85–91
  BECC, 91
  Bridging Boarders Conference of 1994, 86–87, 97, 99
  conferences and meetings, 85–86
  economic analysis, 89–90
  health issues, 89
  highway expansion, 90
  Juntos program, 88–91
  mining, 90
  political issues, 90
  regional interpretive center, 91
  regional profile document, 88
  regional resource inventory, 88
  Roots-Raices-Ta:tk Youth Group, 89
Programme on Protected Areas, 32
Propst, Luther, 69–71, 74, 80, 81, 84, 90, 100, 102, 113, 117–18
Puerto Peñasco, Mexico, 57, 109, 120
Pulliam, H. Ronald, 104

Q

Quabbin to Cardigan Collaborative (Q2C), 206
Quammen, David, 163
Quinn, Daniel, 180
Quintana Silver, Angel, 80, 81, 85, 95

R

Ramon, Tony, 99, 103
Rasker, Ray, 144, 156, 163–64, 188, 190, 192, 199, 202
Reagan, Ronald, 58
Reardon, Carol, 43
Rebert, Paula, 2
Reeves, Brian O. K., 141, 150
Regan, Helen M., 5
Regionalism, 33
Regional Plan Association of America, 33
Reiger, John F., 8, 142
Report on Treaties, Agreements, and Accords . . . , 68
Republican Party, 110
Reyers, Belinda, 28
Reyes-Castillo, Pedro, 60
Rikoon, J. Sanford, 38, 110, 111
Rincon Institute, 69

Rio Grande, 2
Ripley, J. Douglas, 57
Robbins, Jim, 168
Robinson, Bart, 158, 160, 161, 166, 168, 170, 178, 184, 187–88, 191, 194, 196
Rockies Prosperity Act of 2005, 152
Rockwell, David B., 139
Rocky Mountain Ecosystem Coalition, 152, 156
Rocky Mountain National Park and Biosphere Reserve, 99, 102
Rocky Mountains, 136, 139–40
Rodriguez, Emilia, 67
Rogers, Peter J., 28
Rolston, Holmes, 8
Roosevelt, Franklin D., 56, 57
Roosevelt, Theodore, 19
Roots-Raices-Ta:tk Youth Group, 89
Rosenthal, Joshua, 8
Rotary Clubs, 21
Ruggiero, Michael A., 223
Russell, Charlie, 150
Russell, Diane, 29
Russell-Towe, Beth, 149–50
Rüster, Bernd, 20

S

Saguaro National Monument, 62
Saguaro National Park, 70
Salafsky, Nick, 224, 225
Sale, Kirkpatrick, 33–34
Salinas de Gortari, Carlos, 91, 93, 94–95
San Dieguito Complex, 66
Sand Papago people, 66, 67, 72
Sandwith, Trevor, 16, 23–27, 30, 172
Santa Anna, *Serenissima Altera,* 56
Sanwekwe, 22
Save-the-Cedar League (STCL), 150
"Save the whales," 5
Sawchuck, Wayne, 143–44, 150, 152, 189, 197–200
Sawyer, Mike, 152–53, 156, 171, 183, 191, 199

Schaller, George, 22
Schneider, Howard, 155
Schullery, Paul, 144
Schumacher, Robert W., 57, 83–84, 113
Science, of conservation biology, 145–47, 193–97
Scott, Michael, 161, 188
Scott, Peter Markham, 22
Sectionalism, 33
Senge, Peter, 178–79
Sense of Place, A (Harvey), 193
Service, Robert, 176, 177
Seville Biosphere Reserve conference, 105–106, 108
Shafer, Craig L., 39
Shelford, Victor E., 39, 139
Shell, Hanna Rose, 217
Sheridan, Thomas E., 54, 66
Sherrington, Peter, 140
Shine, Clare, 34, 40, 60, 111
Sierra Club, 98, 103, 156
Sierra del Pinacate, 61–68
Sifford, Belinda, 43, 115, 119, 149
Simma, Bruno, 20
Simonian, Lane, 60, 61
VI Symposium on the Gulf of California Environment, 64
Skeele, Tom, 166–67, 184, 188
Slovak National Natural Reserve, 21
Smith, Ceal, 42, 105–106
  demise of the biosphere reserve effort and, 108–109
Smith, George, 152, 197, 199
Smith, Harold, 84, 99–104, 113
Smith, Ted, 159
Smithsonian Institute, 58
Snyder, Gary, 147
Snyder, James D., 226
Sochaczewski, Paul Spencer, 21
"Social epidemics," 165–66, 203
Soles, Roger E., 59, 111
Sonoran Desert, 53–54, 118
  see also International Sonoran Desert Alliance (ISDA)

Sonoran Desert Biosphere Reserve, 64, 103, 105, 113
Sonoran Desert Ecosystem Council, 115
Sonoran Desert Ecosystem Partnership (SDEP), 115–16, 118
Sonoran Desert National Monument (SDNM), 119
Sonoran Desert National Park (SDNP), 119
Sonoran Desert Taskforce, 87, 99–100, 103–106, 113, 118, 231
Sonoran Institute (SI), 65, 67, 88, 89, 90, 113, 114, 156, 171, 188, 199, 202
  Border Land Use Forum of 1992 and, 69–75, 78
  Cooperative Agreement with ORPI, 85, 101
  ISDA and, 79–80, 108
Soulé, Michael, 62, 147, 159
South African Peace Parks Foundation, 32
Species extinction, 5–7, 139
Spencer, Robert, 19
Steffens, Ron, 87, 107
Stegner, Wallace, 3, 145, 187
Steinhart, Peter, 64, 92
Stephens, Doug, 206
Stewart, Bob, 21
Stewart, Craig, 156, 186, 192
Stockholm Conference on the Human Environment of 1972, 23, 37
Stone, Richard, 35
Stroud, Richard H., 217
Struzik, Ed, 164
Successful Communities, 69
Sunstein, Cass R., 166
Sustainable development, 28–29
Suzuki, David, 165, 171
Symington, Fife, 91, 112, 113
Symposium on the Pinacate Ecological Area of 1988, 65, 67–68

T
Tabor, Gary, 28, 140, 141, 146, 152, 153, 159, 166, 185, 191–92, 206
Taft, Tracy, 121

Takacs, David, 7, 8
Tangley, Laura, 39, 41, 58
Territories of chance, 1–3
  conservation effectiveness of, *see* Conservation effectiveness in the territories of change
Tet'it Gwich'in First Nation, 203
Thorsell, Jim, 24
Tijuana River Watershed Project, 14
"Tipping point" theory, 165–66, 203
Tobin, Richard J., 8
Tocqueville, Alexis de, 237–38
Tohono O'odham Nation (TON), 56, 65–68, 72, 86, 87, 98, 101, 105, 111, 230
  on ISDA's board, 82–83
  Pinacate Biosphere Reserve and, 91, 95–96
Tohono O'odham Reservation, 56, 78
Tolentino, Amado S., Jr., 23, 24, 25, 31
Torgrimson, Christine, 171, 190, 192, 204
Toronto *Globe and Mail*, 171
"Tortilla Curtain," 76–78
Trail of the Great Bear (TGB), 149–50
Train, Russell, 36
Transborder conservation, 14–52
  bioregionalism, 33–44
  definition of, 15–17
  migratory species and, 18–20
  new forms of, 43–44
  in North America, 14–15
  peace parks, 20–33
  snapshot of, 17–20
  terminology, 15–16
  transboundary protected areas, 23–33
  water and, 17–18
Transboundary ecosystem management, 15–16
Transboundary natural resource management, 15–16

260 INDEX

Transboundary Protected Areas Task Force (TPATF), 26, 32–33
Transboundary protected areas (TBPA), 23–33
   certification system for, 31
   defined, 23–34
   international support for, 31–33
   levels of cooperation and, 25–26
   pros, cons, and best practices of, 26–31
   types of, 24–25
Transboundary Sonora Desert Letter, 113
Transfrontier Conservation Area (TFCA), 33
Transnational landscapes, 33
Treaty of 1818, 2
Treaty of 1846, 2
Treaty of Alliance of 1780, 20
Treaty of Guadelupe Hidalgo of 1848, 55–56
Trudeau, Pierre, 14
Tucson, Arizona, 77
Tufts University, xiii
Turk Boyer, Peggy, 95
Turner, Anthony M., 223
Turner, Tom, 110

U

Udall, Morris, 103
Udall, Stewart, 63, 103–104, 119
Udvardy, Miklos D. F., 39
Umbrella principle, 146
Unfragmented biodiversity ecosystem coalitions (UBECs), 150–51
United Nations:
   Commission for Europe, 16
   hostility toward, 110–11, 112, 117, 174–75
United Nations Economic and Social Council (ECOSOC), 35
United Nations Educational, Scientific, and Cultural Organization (UNESCO), 34–37, 39, 43, 60, 61, 74, 110, 175

Seville Biosphere Reserve conference, 105–106
U.S. withdrawal from, 58–59
U.S. Agency for International Development, 225
U.S. Border Patrol, 83
U.S. Bureau of Land Management (BLM), 53, 71, 72, 83, 119, 120, 141
U.S. Congress, 110, 112, 139, 152
U.S. Congressional Research Service, 111
U.S. Department of Agriculture (USDA), 57
U.S. Department of Defense (DOD), 57, 71
U.S. Department of State, 112
U.S. Department of the Interior (DOI), 57, 114, 115, 116, 118
U.S. Fish and Wildlife Service (FWS), 19, 20, 57, 139, 141
U.S. Forest Service (USFS), 141, 1442
U.S. Immigration and Naturalization Service (INS), 83
U.S. Marine Corps, 83
U.S.-Mexico Border Field Coordinating Committee, 114, 116
U.S.-Mexico Border States Conference on Recreation, Parks, and Wildlife of 1997 and 1998, 115, 116
U.S.-Mexico Border XXI Program, 4
U.S.-Mexico Convention for the Protection of Migratory Birds and Game Mammals of 1936, 20
U.S.-Mexico Letter of Intent (LOI), 114–16, 118
U.S.-Mexico War, 55
U.S. National Committee for the MAB Program, 58–59
U.S. National Park Service (NPS), 54, 56, 83, 115, 141, 149, 172

MAB Program and, 58, 103–104, 107
U.S. National Science Foundation (NSF), 35
Universidad Nacional Autónoma de México, 92
University for Peace, 32
University of Montana, 151
Upper Gulf of California and Colorado River Delta Bio-sphere Reserve, 91, 92, 93, 109. 115

V

Valdez, Carlos, 87
Valentine, Fernando, 105
Van Tighem, Kevin, 160, 161, 166, 172, 185
Varady, Robert G., 18
Varley, John, 168–69
Vedder, Amy, 22
Vesilind, Priit J., 66, 78, 107
Virgin of Guadalupe, 94
Virunga National Park, 22
Vista, 80–81
Vogel, Gretchen, 35
Volcanos National Park, 22

W

Waldman, Carl, 66
Waldt, Ralph, 149
Walker, Marisa Paula, 76
Walker, Steve, 62
Walking the Big Wild: From Yellowstone to the Yukon on the Grizzly Bears' Trail (Heuer), 164
Waller, Geoffrey, 5
Wapner, Paul, 234, 236
Ward, Peter, 8
Warner, Gerry, 164
Washington Post, 171
Waterton Biosphere Reserve, 43
Waterton-Glacier International Peace Park, 115, 142, 188
Waterton Lakes National Park, 21, 142, 149, 155
Webb, Trish, 153, 161–62
Weber, William, 22
Weisman, Alan, 2, 149
Western Pima County Community Council (WPCCC), 70

INDEX

Western Sonoran Desert Border Committee, 79
Westing, Arthur H., 3, 17, 21, 23, 29, 30, 31
West Wing, The, 172
White, Gilbert, 36
Whittlesey, Lee H., 144
Wiken, Ed B., 223
Wilburforce Foundation, 158, 159, 192, 194
Wilcher, Marshall E., 16
WildCanada, 157
Wild Earth, 147–48
Wilderness Society, 147, 157, 163–64, 171
Wild Hunters, 146
Wildlands Project, The (TWP), 87, 147–48, 157
Wildlife, 56, 135–36, 139–53
  connectivity concept, 143–47, 167, 178–82
  Hornaday and, 217–19
Wild Life Conservaton in Theory and Practice (Hornaday), 218
Wild Rockies Earth First! (WREF), 151
WildSight, 157, 171
Wilkie, D. S., 22
Wilkinson, Todd, 198
Willcox, Louisa, 139, 140, 152, 156, 187, 188–89
Williams, Florence, 56, 61, 62, 67, 83, 107
Williams, Ted, 20
Willmore Wilderness Park, 143
Wilson Edward O., 6, 7, 220
Wise, Timothy A., 238
Wise-use movement, 181, 182
Wolf, Aaron T., 18
Wolf, Alan, 165
Wolke, Howie, 151
Wolves, 146, 163, 167
Wood, Alexander, 220, 221
World Bank, 32
World civil politics, 236
World Commission on Protected Areas (WCPA), 30, 32
World Conservation Monitoring Center, 223
World Conservation Union (IUCN), 5, 22, 41
IBP and, 35
TBPAs and, 23, 24, 32–33
World Heritage Convention, 110
World Resources Institute, 225
World Wildlife Fund (WWF), 32, 69, 225, 236
  -Canada, 146, 148, 166
Worthington, E. Barton, 35
Wuerthner, George, 141, 142, 144, 168, 198

Y

Yale University, 152, 193
Yellowstone Center for Resources, 168
Yellowstone National Park, xiii, 22, 110, 142, 144, 162, 166–69, 177, 178, 193
Yellowstone Park Timber Land Reserve, 142
Yellowstone to Yukon: Canadian Conservation Efforts and a Continental Landscape/Bio-diversity Strategy, 159
Yellowstone to Yukon Conservation Initiative: Fiction vs. Fact, The, 163
Yellowstone to Yukon Conservation Initiative (Y2Y), xiii–xv, 34, 134–216
  Aboriginal Advisory Group, 173
  "Atlas" of, 155, 193
  "big landscapes" and, 166–67, 178–82
  Biodiversity Strategy Communications Plan, 153–55, 161–64
  biosphere reserves and land-use patterns, 39–42
  borders within, 140–41
  branding of, 204
  civil society and, 4
  as a coherent ecological unit, 139
  community sustainability and, 160
  conceptual foundations of, 4
  conflict within, 156, 190–91
  conservation biology and, 145–46, 147, 193–97
  conservation effectiveness of, *see* Conservation effectiveness in the territories of change
  conservation organizations and, 147–48
  conservation successes of, *see* Conservation successes of Y2Y
  constituencies of, 162, 230–31
  Coordinating Committee (CC)/board, 155–60, 175, 185, 188, 190, 193, 194, 205, 231
  core areas in, 142
  Council/partners, 155–58, 190
  described, 136–37
  early issues faced by, 153–55
  effectiveness of, 165
  empowerment and, 237
  expanding the scale of conservation for conservationists, 164–68
  first conference of, 155–56
  foundation supporting, 158–59
  general public and, 170–72
  geographic predecessors of, 168–70
  goals of, 134
  government and, 172–74
  iconic composition of, 176–82
  indigenous cultures and, 141, 173, 198, 202–203, 230–31
  institutional development of, 153–59
  Kananaskis meetings and, 144, 148, 153, 155, 158
  Locke and, 142–49, 153
  map of, 137, 138
  media and, 171–72, 174
  membership in, 158
  mission of, 159–64, 230

name changes, 154
as new form of transborder conservation, 43–44
"Northern Rockies" and, 139–40, 197
opposition to, 164, 166, 167–68, 174–76
origins of, 141–53
publications of, 163
region encompassed by, 137–41
river systems, 140
scientific research and, 146–47, 193–97
staff of, 158, 162
threats to biodiversity, 140
today and tomorrow, 203–206
U.S. states vs. Canadian provinces, 141
vision, 178–82
vision backlash, 174–76
web of conception, 141–53, 154
wildlife and, 135–36, 139–53, 167
Y2Y Inc., 153–59
Yellowstone to Yukon (NGS), 171
Yetman, David, 61, 62, 63, 77
Yoho National Park, 142
Young, Don, 112
Young, Oran, 226–28, 229
Yruretagoyena, Carlos, 116
Y2Y Conservation News, 163
Y2Y Conservation Science Grants, 194
Y2Y Hike, 164, 172
Y2Y Q&A 2001, 163
Yukon, 200–203

Z

Zakin, Susan, 62, 78, 112, 114, 115, 152
Zammuto, Rick, 150
Zbicz, Dororthy, 24, 25–26
Zedillo, Ernesto, 114
Zeller, Tom, 76

ISLAND PRESS BOARD OF DIRECTORS

Victor M. Sher, Esq. (*Chair*)
Sher & Leff
San Francisco, CA

Dane A. Nichols (*Vice-Chair*)
Washington, DC

Carolyn Peachey (*Secretary*)
Campbell, Peachey & Associates
Washington, DC

Drummond Pike (*Treasurer*)
President
The Tides Foundation
San Francisco, CA

David C. Cole
Chairman, President, and CEO
Maui Land & Pineapple Company, Inc.
Kahului, Maui, HI

Catherine M. Conover
Quercus LLC
Washington, DC

Merloyd Ludington Lawrence
Merloyd Lawrence Inc.
Boston, MA

William H. Meadows
President
The Wilderness Society
Washington, DC

Henry Reath
Princeton, NJ

Will Rogers
President
The Trust for Public Land
San Francisco, CA

Alexis G. Sant
Trustee and Treasurer
Summit Foundation
Washington, DC

Charles C. Savitt
President
Island Press
Washington, DC

Susan E. Sechler
Senior Advisor
The German Marshall Fund
Washington, DC

Peter R. Stein
General Partner
LTC Conservation Advisory Services
The Lyme Timber Company
Hanover, NH

Diana Wall, Ph.D.
Director and Professor
Natural Resource Ecology Laboratory
Colorado State University
Fort Collins, CO

Wren Wirth
Washington, DC